FROM SCIENCE TO BUSINESS

Preparing Female Scientists and Engineers for Successful Transitions into Entrepreneurship

Summary of a Workshop

Catherine Jay Didion, Rita S. Guenther, and Victoria Gunderson, *Rapporteurs*

Committee on Women in Science, Engineering, and Medicine

Policy and Global Affairs

NATIONAL RESEARCH COUNCIL
OF THE NATIONAL ACADEMIES

NATIONAL ACADEMIES PRESS
Washington, D.C.
www.nap.edu

THE NATIONAL ACADEMIES PRESS   500 Fifth Street, N.W.   Washington, DC 20001

NOTICE: The project that is the subject of this report was approved by the Governing Board of the National Research Council, whose members are drawn from the councils of the National Academy of Sciences, the National Academy of Engineering, and the Institute of Medicine. The members of the committee responsible for the report were chosen for their special competences and with regard for appropriate balance.

This study was supported by the National Academy of Sciences. Any opinions, findings, conclusions, or recommendations expressed in this publication are those of the author(s) and do not necessarily reflect the views of the organizations or agencies that provided support for the project.

International Standard Book Number-13: 978-0-309-25609-4
International Standard Book Number-10: 0-309-25609-7

Additional copies of this report are available from the National Academies Press, 500 Fifth Street, N.W., Keck 360, Washington, DC 20001; (800) 624-6242 or (202) 334-3313 Internet, *http://www.nap.edu*

Copyright 2012 by the National Academy of Sciences. All rights reserved.

Printed in the United States of America.

THE NATIONAL ACADEMIES
Advisers to the Nation on Science, Engineering, and Medicine

The **National Academy of Sciences** is a private, nonprofit, self-perpetuating society of distinguished scholars engaged in scientific and engineering research, dedicated to the furtherance of science and technology and to their use for the general welfare. Upon the authority of the charter granted to it by the Congress in 1863, the Academy has a mandate that requires it to advise the federal government on scientific and technical matters. Dr. Ralph J. Cicerone is president of the National Academy of Sciences.

The **National Academy of Engineering** was established in 1964, under the charter of the National Academy of Sciences, as a parallel organization of outstanding engineers. It is autonomous in its administration and in the selection of its members, sharing with the National Academy of Sciences the responsibility for advising the federal government. The National Academy of Engineering also sponsors engineering programs aimed at meeting national needs, encourages education and research, and recognizes the superior achievements of engineers. Dr. Charles M. Vest is president of the National Academy of Engineering.

The **Institute of Medicine** was established in 1970 by the National Academy of Sciences to secure the services of eminent members of appropriate professions in the examination of policy matters pertaining to the health of the public. The Institute acts under the responsibility given to the National Academy of Sciences by its congressional charter to be an adviser to the federal government and, upon its own initiative, to identify issues of medical care, research, and education. Dr. Harvey V. Fineberg is president of the Institute of Medicine.

The **National Research Council** was organized by the National Academy of Sciences in 1916 to associate the broad community of science and technology with the Academy's purposes of furthering knowledge and advising the federal government. Functioning in accordance with general policies determined by the Academy, the Council has become the principal operating agency of both the National Academy of Sciences and the National Academy of Engineering in providing services to the government, the public, and the scientific and engineering communities. The Council is administered jointly by both Academies and the Institute of Medicine. Dr. Ralph J. Cicerone and Dr. Charles M. Vest are chair and vice chair, respectively, of the National Research Council.

www.national-academies.org

COMMITTEE FOR FROM SCIENCE TO BUSINESS: PREPARING FEMALE SCIENTISTS AND ENGINEERS FOR SUCESSFUL TRANSITIONS INTO ENTREPRENEURSHIP WORKSHOP

LILIAN WU, *Chair*, Program Executive, IBM Corporation
ALICE AGOGINO (NAE),* Roscoe and Elizabeth Hughes Professor of Mechanical Engineering, University of California, Berkeley
ALLAN FISHER, Senior Vice President, Laureate Education, Inc.
SHELDON SCHUSTER, President, Keck Graduate Institute of Applied Life Science
LYDIA VILLA-KOMAROFF, Chief Scientific Officer, Cytonome/ST, LLC
SUSAN WESSLER (NAS),* Distinguished Professor of Genetics, University of California, Riverside

STAFF

CATHERINE DIDION, Director
RITA S. GUENTHER, Program Officer
WEI JING, Research Associate

*Denotes members of the National Academy of Science (NAS), National Academy of Engineering (NAE), and Institute of Medicine (IOM).

COMMITTEE ON WOMEN IN SCIENCE, ENGINEERING, AND MEDICINE

POLICY AND GLOBAL AFFAIRS NATIONAL RESEARCH COUNCIL

RITA R. COLWELL (NAS),* *Chair*, Distinguished Professor, University of Maryland, College Park and Bloomberg School of Public Health, Johns Hopkins University
ALICE AGOGINO (NAE),* Roscoe and Elizabeth Hughes Professor of Mechanical Engineering, University of California, Berkeley
JOAN W. BENNETT (NAS),* Professor of Plant Biology and Pathology, Rutgers University
JEREMY M. BERG (IOM), Associate Senior Vice Chancellor for Science, University of Pittsburgh
VIVIAN PINN (IOM),* Director for Research on Women's Health, National Institutes of Health, Emeritus
PATRICIA TABOADA-SERRANO, Assistant Professor, Department of Chemical and Biomedical Engineering, Rochester Institute of Technology
LYDIA VILLA-KOMAROFF, Chief Scientific Officer, Cytonome/ST, LLC
SUSAN WESSLER (NAS),* Distinguished Professor of Genetics, University of California, Riverside

STAFF

CATHERINE DIDION, Director
RITA S. GUENTHER, Program Officer
WEI JING, Research Associate

*Denotes members of the National Academy of Science (NAS), National Academy of Engineering (NAE), and Institute of Medicine (IOM).

PREFACE AND ACKNOWLEDGMENTS

Scientists, engineers, and medical professionals play a vital role in building the 21st century science and technology enterprises that will create solutions and jobs critical to solving the large, complex, and interdisciplinary problems faced by society—problems in energy, sustainability, the environment, water, food, disease, and healthcare. As a growing percentage of the scientific and technological workforce, women need to participate fully not just in finding solutions to technical problems, but also in building the organizations responsible for the job creation that will bring these solutions to market and to bear on pressing issues. To accomplish this, it is important that more women in science and engineering become entrepreneurs in order to start new companies; create business units inside established organizations, mature companies, and the government; and/or function as social entrepreneurs focused on societal issues.

Entrepreneurship represents a vital source of change in all facets of society, empowering individuals to seek opportunity where others see insurmountable problems. Technology entrepreneurship as a style of business leadership involves identifying high-potential, technology-intensive commercial opportunities, gathering resources such as talent and capital, and managing rapid growth and significant risks using principled decision-making skills.

There is concern among experts that women are not adequately applying their technical training to entrepreneurship, certainly not at the same rate as men. We should be aware that in addition to educating women scientists and engineers in rigorous problem solving, it is equally important to provide exposure and training to impart the skills that will enable more women to move from the role of expert to that of leader in dynamic new business enterprises. As one workshop participant noted, women-owned businesses accounted for 40 percent of all privately-owned businesses in 2008, contributing $3 trillion to the U.S. economy.[1] Women scientists and engineers are also needed to organize, create, and manage ventures in the not-for-profit world to effect social change. "Social entrepreneurship" creates ventures that interweave money-generation with the furtherance of social and environmental goals.

In August 2009, the Committee on Women in Science, Engineering, and Medicine (CWSEM), convened a workshop entitled, "From Science to Business: Preparing Female Scientists and Engineers for Successful Transitions into Entrepreneurship" to assess the current status of women undertaking entrepreneurial activity in technical fields, to better understand the nature of the barriers they encounter, and to identify what it takes for women scientists and engineers to succeed as entrepreneurs. The workshop focused on women's career transitions from academic science and engineering to entrepreneurship, with a goal of identifying knowledge gaps in women's skills as well as experiences crucial to future success in business and critical for achieving leadership positions in entrepreneurial organizations. More than two-thirds of workshop participants were from academia, with the remainder from industry, non-

---

[1] Susan Windham Bannister. "Bridging the Gaps: Entrepreneurship, Science, and Gender." Presented at the workshop, August 31, 2009.

profit organizations, independent consulting, and elsewhere. Among the academics, over 60 percent of the participants identified themselves as graduate students or postdoctoral researchers.

This topic is a continuation of an earlier CWSEM workshop titled "From Doctorate to Dean or Director: Sustaining Women through Critical Transition Points in Science, Engineering, and Medicine," which examined critical transition points in both academia and industry, and how to strengthen women's participation and advancement during the process. CWSEM serves as a focal point on gender for the three academies: National Academy of Sciences, National Academy of Engineering, and Institute of Medicine. This workshop was an impetus for CWSEM to determine in 2011 that one of its focal points should be on innovation, job creation and the role of entrepreneurship in developing scientific career pathways.

Over the course of the workshop that began on August 31, 2009, a wide variety of studies, panels, and reports demonstrated that the problem of preparing women scientists and engineers for successful transitions into entrepreneurship is multi-faceted. Technical entrepreneurship is of interest to women—and particularly young researchers—but access to entrepreneurial training and information resources is often lacking in home institutions.

This summary provides an overview of the individual presentations and panel discussions at the workshop, in the order in which they were presented. It has been prepared by the workshop rapporteurs as a factual summary of what occurred at the workshop. The planning committee's role was limited to planning and convening the workshop. The statements made in this summary are those of the rapporteurs or individual workshop participants and do not necessarily represent the views of all of the workshop participants, the planning committee, CWSEM, or the National Academies.

This report has been reviewed in draft form by individuals chosen for their diverse perspectives and technical expertise, in accordance with procedures approved by the National Academies' Report Review Committee. The purpose of this independent review is to provide candid and critical comments that will assist the institution in making its published report as sound as possible and to ensure that the report meets institutional standards for objectivity, evidence, and responsiveness to the study charge. The review comments and draft manuscript remain confidential to protect the integrity of the process.

We wish to thank the following individuals for their review of this report: Joanne McGrath Cohoon, National Center for Women and Information Technology; Baat Enosh, University of Colorado; Mary Juhas, Ohio State University; Karla Shepard Rubinger, Rosalind Franklin Society; and Carolyn Vallas, University of Virginia. Although the reviewers listed above have provided many constructive comments and suggestions, they were not asked to endorse the content of the report, nor did they see the final draft before its release. Responsibility for the final content of this report rests entirely with the rapporteurs and the institution.

> Lilian Wu, Chair
> Committee on Science to Business: Preparing Female
> Scientists and Engineers for Successful Transitions into
> Entrepreneurship

CONTENTS

1 Entrepreneurial Careers of Women ... 1
E. J. Reedy, Manager, Research and Policy, Kauffman Foundation ... 1

2 Panel I: From Bench to Business: Career Paths for Ph.D.s ... 5
Laurel Smith-Doerr, Associate Professor of Sociology, Boston University ... 5
Lydia Villa-Komaroff, Chief Scientific Officer, Cytonome/ST, LLC ... 8
Susan Windham-Bannister, President and CEO, Massachusetts Life Science Center ... 10

3 Panel II: Aspects of Leadership in Biotechnology Careers ... 13
Judy Heyboer, Human Resources Consultant, Former Senior Vice President, Genentech, Inc. ... 13
Barbara Wallner, President and CEO, Chymic Therapeutics, Inc. ... 14
Lydia Villa-Komaroff, Chief Scientific Officer, Cytonome/ST, LLC ... 15

4 Panel III: Education to Prepare for Entrepreneurial Careers ... 17
Sheldon M. Schuster, President, Keck Graduate Institute of Applied Life Sciences ... 17
Gail Naughton, Dean, College of Business, San Diego State University and Founder, Advance Tissue Science, Inc. ... 18
Michael Teitelbaum, Program Director, Alfred P. Sloan Foundation ... 19
Jessica Townsend, Assistant Professor, Mechanical Engineering, Olin College ... 21

5 Studies on Entrepreneurship ... 23
Caroline Simard, Director of Research and Executive Programs, Anita Borg Institute for Women and Technology ... 23
Manwai (Candy) Ku, Researcher, Stanford University ... 26

6 Panel IV: Alternative Forms of Entrepreneurships in Sustainable Technologies: Intrapreneurship in Corporations and Government, Social Entrepreneurship, and Traditional Entrepreneurship ... 29
Sharon Nunes, Vice President, IBM Green Innovations ... 29
Maxine L. Savitz, General Manager for Technology Partnerships, Honeywell Inc. (retired) and former Deputy Assistant Secretary for Conservation, U.S. Department of Energy ... 30
Judith Giordan, Senior Advisor, National Collegiate Innovators and Inventors Alliance ... 30
Lucinda Sanders, CEO and Co-founder, National Center for Women & Information Technology ... 31

7 Themes from the Workshop and Closing Remarks ... 33

Appendix A Workshop Agenda ... 35

Appendix B Committee on Women in Science, Engineering and Medicine: 39
Member Biographies

Appendix C Speakers Biographies   43

Appendix D Workshop Participants   51

1

ENTREPRENEURIAL CAREERS OF WOMEN

E. J. Reedy, Manager, Research and Policy, Kauffman Foundation

E. J. Reedy, Manager in Research and Policy at The Ewing Marion Kauffman Foundation

The opening workshop address was designed to provide an overview of women's status and participation in entrepreneurial careers. E. J. Reedy presented findings from a Kauffman Foundation study titled "Sources of Financing for New Technology Firms: A Comparison by Gender" (2009)[1] and noted that a relatively small number of studies have specifically examined the experience of women in high-tech entrepreneurship.[2] The Kauffman Foundation study shows that among all surveyed startup firms, only 15 percent of those in the biotechnology and high-technology sectors reported having a female primary owner as compared to 30 percent female-owned startup firms in all other sectors. In the initial start-up year, among the high-tech start-ups examined, 20 percent of male-owned firms reported obtaining formal equity (primarily venture capital or angel funding), compared to 7 percent of female-owned firms. Reddy suggested that this occurs because the venture capital industry is relatively closed and male-dominated. His studies show that women-owned firms represent a small minority of the overall venture capital backed firms, which may contribute to the lack of initial formal equity. He noted that an

> *"When we put a gender lens on where the barriers are, it is around growth opportunities."*
>
> E. J. Reedy, Manager, Research and Policy, The Ewing Marion Kauffman Foundation

---

[1] Robb, A. and S. Coleman, (2009), Sources of Financing for New Technology Firms: A Comparison by Gender: Fifth in a series of reports using data from the Kauffman Firm Survey. Retrieved August 15, 2009, from http://www.kauffman.org/uploadedFiles/ResearchAndPolicy/Sources%20of%20Financing%20for%20New%20Technology%20Firms.pdf.

[2] Since this August 2009 workshop, "From Science to Business," the Kauffman Foundation has produced the following reports on scientific women in entrepreneurship:

Robb, A. and Coleman, S. (2009,), "Characteristics of New Firms: A Comparison by Gender," The Kauffman Foundation. Retrieved from http://www.kauffman.org/uploadedFiles/kfs_gender_020209.pdf.;

Cohoon, J. M., V. Wadhwa, and Mitchell L. (2010). "Are Successful Women Entrepreneurs Different than Men?" [Working Paper] Retrieved from http://papers.ssrn.com/sol3/papers.cfm?abstract_id=1604653, and

Mitchell, L. (2011), "Overcoming the Gender Gap: Women Entrepreneurs as Economic Drivers," The Kauffman Foundation. Retrieved from
http://www.kauffman.org/uploadedFiles/Growing_the_Economy_Women_Entrepreneurs.pdf.

additional factor may be that female owners of high-technology firms have less overall managerial experience than their male counterparts and are less likely to have previously owned start-up companies in the same field, since 55 percent of male owners have previously owned a high-technology start-up compared to 12 percent of females. Reedy pointed out that venture capitalists view this serial behavior—and even serial failure—as a mark of an entrepreneur being tested, because investors value an entrepreneur's willingness to start over again and again.

Reedy further elaborated on the gender gap in the high-tech industry by quoting similar findings by other researchers. A 2006 study by Cross and Linehan found that in established high-tech organizations, women are often excluded from formal and informal networks that would otherwise provide access to managerial or technical leadership positions in those firms.[3] Similarly, a 2005 study by Tai and Randi L. Sims[4] found that women had difficulty gaining senior management experience that would make them attractive to external capital providers, should they start their own companies. Reedy considered the pervasive culture in such enterprises as a plausible reason for this gap, where issues such as work-life balance and family-friendly policies affect gender equity.

Despite the existing gaps, Reedy noted that women-owned firms represent a growing component of the small-business sector. According to a U.S. Census Bureau report,[5] there were 6.5 million privately held women-owned firms in the United States in 2002. These firms generated an estimated $940 billion in sales and employed 7.1 million people. He continued to comment that women-owned firms now account for 30 percent of all firms, including self-employment and larger businesses. From 1997 to 2002, the number of women-owned businesses increased 20 percent. However, he noted that revenues from women-owned businesses increased less than 15 percent during this period, compared to a 22 percent revenue increase for all businesses.

A study on women-owned high-tech firms in four metropolitan regions in the United States—Silicon Valley, Boston, Washington, D.C., and Portland, Oregon— found that in all four regions women-owned high-tech firms were smaller in terms of average revenue and employment.[6] Reedy suggested that women are more likely to form companies alone than in partnership with men, even though studies indicate that the success rate for new companies increases as the number of company founders increases. Women entrepreneurs are also more likely to participate in high-tech sectors such as software publishing, computer systems design services, research services, and management and consulting services; whereas men are more likely to establish companies in the manufacturing sector. Interestingly, Reedy noted both male- and female-owned start-ups were initially shown to be based out of the home, 50 percent of the time for males and 60 percent of the time for females, suggesting a focus on consulting efforts.

Female-owned high-tech start-ups do show a slightly higher survival rate than those owned by men. However, Reedy noted that women entrepreneurs launch high-technology firms with less financial capital than men, and continue to follow a different financial strategy over

---

[3] Cross, C., & Linehan, M. (2006). Barriers to Advancing Female Careers in the High-tech Sector: Empirical Evidence from Ireland. *Women in Management Review*. 1, 28.
[4] Tai, An-Ju R., and Randi L. Sims (2005). The Perception of the Glass Ceiling in High-Technology Companies. *Journal of Leadership and Organizational Studies*. 12 (1), 16–23.
[5] U.S. Census Bureau (2006). Women-owned Business Grew at Twice the National Average, Census Bureau Reports. Retrieved August 15, 2009 from
http://www.census.gov/newsroom/releases/archives/business_ownership/cb06-14.html.
[6] Mayer, H. (2008), Segmentation and Segregation Patterns of Women-owned High-tech Firms in Four Metropolitan Regions in the United States, *Regional Studies*. 42 (10), 1357-1383.

time. Women's reliance on internal funding sources (e.g., owners' savings, loans from family and friends, credit cards) makes a difference such that years after startup, women-owned high-tech firms lag behind men-owned firms in numerous performance measures, including revenues, assets, and employment. However, profits were shown to be higher for female-owned firms.

In addition, Reedy discussed the age at which entrepreneurs usually start their businesses. One of the common misconceptions about entrepreneurs is that these individuals are age 25 to 30.[7] In fact, it is more common for those in their late 30s and early 40s to start companies because at that age people have considerable industry experience and starting a business is a strategic part of their overall career goals. This average start-up owner age was not shown to vary significantly between men and women or across sectors. But women owners reported working slightly fewer hours per week than their male counterparts.

---

[7] Kauffman Foundation Survey. Available at: http://www.kauffman.org/research-and-policy/kauffman-firm-survey.aspx.

2

PANEL I:
FROM BENCH TO BUSINESS: CAREER PATHS FOR PH.D.S

Laurel Smith-Doerr, Associate Professor of Sociology, Boston University
Lydia Villa-Komaroff, Chief Scientific Officer, Cytonome/ST, LLC
Susan Windham-Bannister, President and CEO, Massachusetts Life Science Center

Laurel Smith-Doerr, Associate Professor of Sociology, Boston University

Laurel Smith-Doerr provided sociological data about women in the life science fields of academia. She further discussed the career contexts for doctoral students in the life sciences, specifically with regard to biotechnology firms that are focused on human therapeutics and diagnostics. She focused on how to prepare entrepreneurship contexts in science and engineering for gender equity in entrepreneurship and innovation. Her sociological analysis of the transition from academic institutions to start-up firms has focused primarily on the life sciences, although over the past few decades there has been significant blurring of the boundaries between academic biology and bio-tech entrepreneurship. In academic institutions, women and men have reached parity at the doctoral student level and similarly, women are also well-represented at post-doc and assistant professor levels at about 45 percent of their total respective populations. However, females still only represent about 30 percent of full-time senior faculty members in the life sciences (see Figure 2-1).

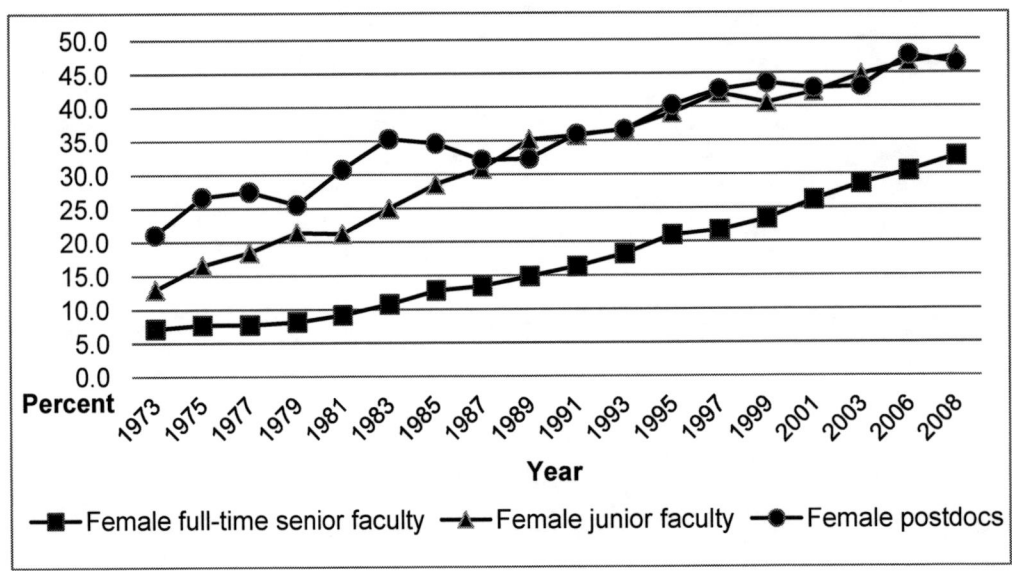

FIGURE 2-1 Percentage of females in academic life science positions, 1973-2008.[1]
SOURCE: National Science Foundation. *Science and Engineering Indicators 2006* and *Science and Engineering Indicators 2012*.

In discussing women's entrepreneurship in the academic life sciences, Smith-Doerr cited Fiona Murray's findings on this topic: "within academic life sciences, women's entrepreneurship is evident in the form of faculty founding companies, patenting, inclusion on scientific advisory boards, and industry co-authorship."[2] She emphasized that these entrepreneurship-driven faculty members tend to be full professors. Therefore, in light of the tenure gap exhibited in Figure 2.1, Smith-Doerr suggested that biotechnology entrepreneurship tends to be male-dominated, so that of those academics who become entrepreneurs, only 4.7 percent of company founders and 5.6 percent of scientific advisory board members are women.

Looking beyond the company founder and scientific advisory board level, Smith-Doerr researched the participation and equity of females in biotechnology firms relative to established multi-national companies and academic environments. In this context, she briefly highlighted the influence of unconscious and implicit biases. Current research demonstrates the need for women to be more productive than men to achieve the same employment stature, as well as the importance of work-life balance. This, then, led her to explicitly focus on the role of organizational structure and its function in understanding the gender gap. Specifically, Smith-Doerr categorized life science organizations as either "hierarchical" organizations, such as multi-national pharmaceutical companies and academic institutions, or "networking" organizations, such as biotechnology, entrepreneur-driven firms. Distinctions between these organizations arise in their communication models and employee interaction procedures. Hierarchical organizations tend to follow strict rules that are strongly influenced by the ranks of the interacting members,

---

[1] Data abstracted from *Science and Engineering Indicator 2006* appendix table 5-23, and *Science and Engineering Indicators 2012* appendix table 5-13. Data points for 2006 and 2008 were added in Figure 2-1 to data presented at the workshop.
[2] Murray, F., Graham, L. (2007). Buying and selling science: gender differences in the market for commercial science, *Industrial and Corporate Change,* 16, 657-689.

compared to networking organizations that follow cultural norms driven by interpersonal relationships and social capital.

Smith-Doerr noted that there is also a significant difference in the prevalence of women and men in leadership positions with hierarchical versus networking organizations. Women are eight times more likely than men to move into supervisory positions in network-structured firms; however, women are significantly less likely than men to move into supervisory roles in hierarchical organizations. She further noted that men showed no difference in their propensity to advance into supervisory roles between organizational structures.[3] Smith-Doerr elaborated on the probabilities of patenting in both organizational structures as shown in Figure 2-2. Only in the networking-type industry settings does gender equity in patenting productivity occur. In all other organizational structures, including industrial hierarchical organizations, significant gender gaps are evident, with increased probabilities for men to patent.

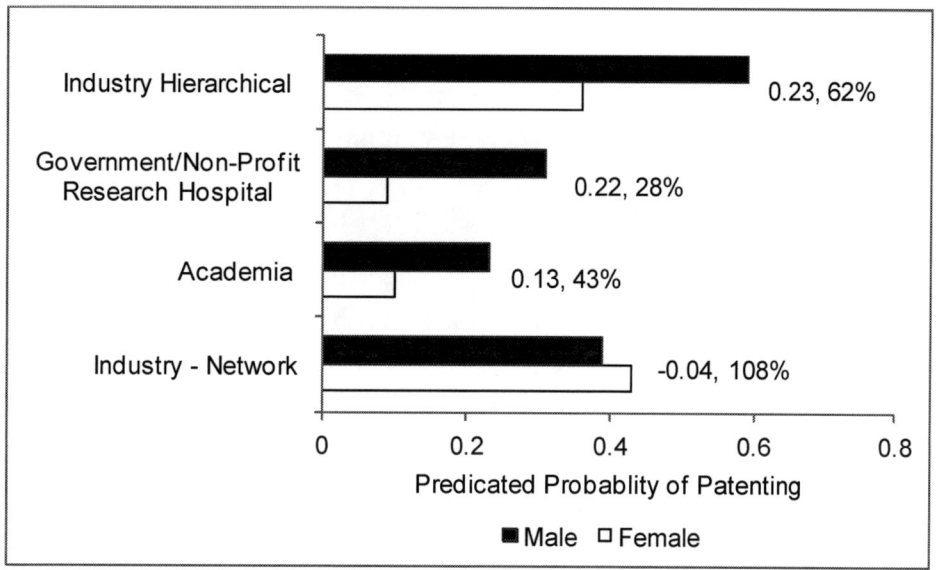

FIGURE 2-2 Predicted probabilities of patenting, by field and gender.
NOTE: Data labels refer to the difference in probabilities between men and women (male-female) and the female/male predicted probability ratio (multiplied by 100). All other variables were held at mean.
SOURCE: Whittington, K.B. and Smith-Doerr, L (2008). Women inventors in context: Disparities in Patenting across Academia and Industry. *Gender and Society.* 22 (2): 194-218.

To understand both the leadership and patenting gender gap trending explained above, Smith-Doerr cited interview studies she and her colleagues have performed. She emphasized three recurrent themes from these interviews that help to explain the greater gender equity at biotechnology firms: flexibility in collaboration, increased organizational transparency, and emphasis on collective rewards. Smith-Doerr further explained that these characteristics are examples of network organizations in which there is indefinite and sequential interaction structure, norms govern relations, partners pool resources, expectations foster collaboration but

---

[3] Smith-Doerr, L. (2004), based on logistic regression analysis controlling for years since Ph.D., prestige of Ph.D. program; N=2,062.

are not rule bound, and there is a non-redundant, "freer" flow of information, which interviews suggest lead to greater gender equity.

Moving beyond biotechnology and looking at the transferability of these findings, Smith-Doerr emphasized that the global challenges that may motivate new entrepreneurship appear to require an interdisciplinary approach. According to National Science Foundation (NSF) data, in 2006, 45 percent of the life scientists employed in business and industry were women and 46 percent of those employed in academia were women. Representation of women physical scientists employed in business and industry was 30 percent—the same percentage as in academia.[4] Despite these wider gender gaps in non-life science fields, Smith-Doerr suggested that interdisciplinary research appears to draw women. She further noted that entrepreneurship is also becoming more interdisciplinary, suggesting the possibility that entrepreneurship may become more gender equal in the future. She suggested that as more women become attracted to the entrepreneurial pipeline, other sectors may begin to follow the trending observed for biotechnology networking organizations.

Lydia Villa-Komaroff, Chief Scientific Officer, Cytonome/ST, LLC

To follow up with Smith-Doerr's discussion of gender equity in biotechnology entrepreneurship, Lydia Villa-Komaroff gave a summary of the European Commission – United States Task Force on Biotechnology Research Workshop "A Global Look at Women's Leadership in Biotechnology Research." This workshop brought together representatives from United States and the European Union funding agencies to develop a mutual understanding of gender diversity in both areas and to develop a series of recommendations and action items that could lead to change.

In both the United States and the European Union, Villa-Komaroff noted that there is a large decline in the representation of women as their academic careers progress. She highlighted this trend with Figure 2-3, dubbed the "scissors diagram," which indicates the large disparity between men and women as they progress to higher levels in their European academic careers. She noted the need for institutional change, such as active recruitment of women to combat this gender inequity. She cited recent data demonstrating that at the doctoral degree granting level, the number of science and engineering degrees awarded to women has increased over the past 30 years, but has recently leveled off. Villa-Komaroff suggested that these leveled numbers may be explained by the hypotheses that European doctoral students are only trained in a limited number of areas that do not include business and/or entrepreneurship training, for example, on the realities of the budgeting process. Therefore, she suggested that in order to continue the success of educating the next generation new models for doctoral degree programs need to arise.

---

[4] National Science Foundation (2010). *Women, Minorities and Persons with Disabilities in Science & Engineering*. Updated from the 2003 data presented at the workshop.

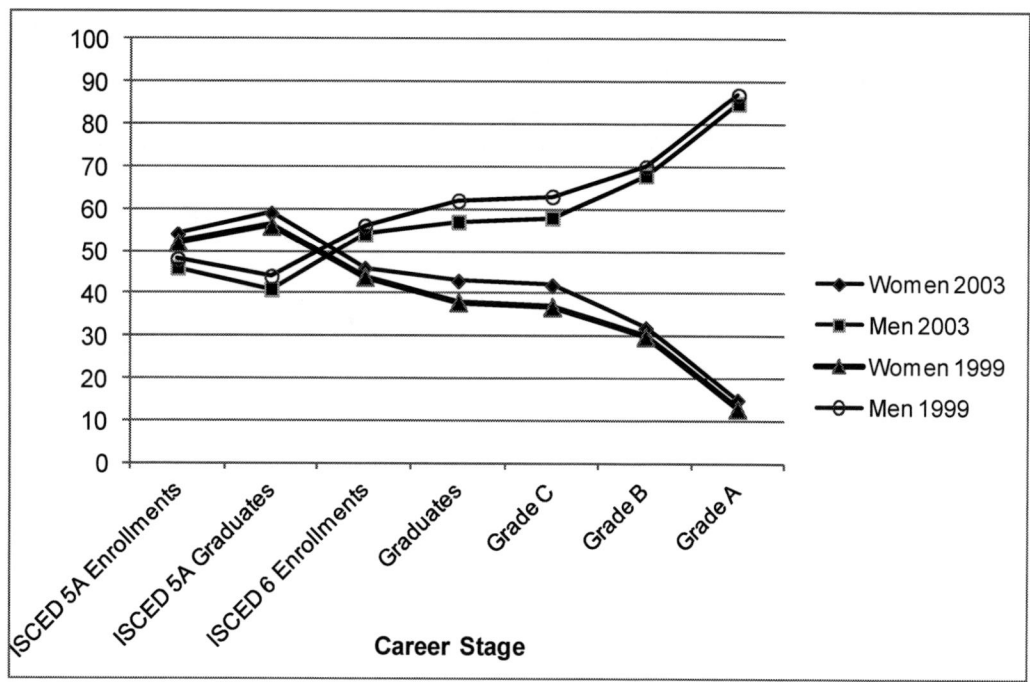

FIGURE 2-3 Proportion of men and women at different stages of a typical academic career, EU 25.
NOTE: The career stages are defined as follows: 5A (bachelors and masters), 6 (advanced higher education programs, doctoral), Grade C (the first grade/post into which a newly qualified doctoral graduate would be recruited), Grade B (researchers working in positions not as senior as top positions [Grade A] but more senior than Grade C), and Grade A (the single highest grade/post at which research is normally conducted).
SOURCE: Eurostat Education Database. Presented by Andrew Collins, University of Olson, Norway at the workshop *A Global Look at Women's Leadership in Biotechnology Research* in June 2009. Adapted from European Commission report *Mapping the Maze: Getting More Women to the Top in Research.*

Furthermore, Villa-Komaroff presented factors that may help women to achieve greater success in science and engineering fields. These factors include:

- Training: Scientific and business training are both necessary. An understanding of business is important, however not inherent in typical doctoral programs.
- Support: In any organizational setting, it is critical to have a supportive environment with mentors at every career stage and a viable network. Female and male networks differ, and some data suggests that female networks are ineffective at advancing women.
- Personal characteristics: Key personal traits for success include persistence, tenacity, flexibility and self-confidence, with additional emphasis on self-confidence and risk-taking.

Villa-Komaroff indicated that policies, intentions, and rules are critical to effectively changing the status-quo. She further noted that institutional changes at the top levels will not induce large-scale transformations; instead efforts to "move down" the educational system are necessary. She also underscored the recommendations from the E.U.-U.S. workshop that included the need for:

- better collection and dissemination of data
- implementation of policies that promote good practices
- achieving buy-in at the highest levels as well as at the grassroots level
- creation of formal and informal mechanisms to provide women with an understanding of what it takes to be successful, i.e. teach them the "rules of the game"
- better accountability of various programs

Susan Windham-Bannister, President and CEO, Massachusetts Life Science Center

Susan Windham-Bannister continued the "business of science" discussion by providing data complementary to that of the previous panelists. She also elaborated on her observations and hypotheses concerning how to bridge the gaps between gender, entrepreneurship, and science.

"What is the story of women in science? It is the story of the few and the fewer," said Windham-Bannister. She noted that while there are more women than men in the top 10 percent of fundable National Institute of Health (NIH) proposals, fewer grants providing fewer dollars are awarded to women. Similar gender gaps can be seen in other avenues of science as well. For instance, less than 3 percent of Nobel Laureates have been women and over the years; women have held only 27 percent of the jobs in science and engineering. She suggested that discriminatory employee practices are visible from the very start of a career in sciences: women earn 24 percent less than their male counterparts.[5] "When you look at women entrepreneurs, it is a somewhat better story but it's not a great story," she said.

Windham-Bannister noted that 28-30 percent of primary owners of privately-held businesses are women, but the percentage of women owners increases to 40 percent when also considering women who are part-owners of privately-held businesses. However, she continued that the performance matrix for female-led new ventures lags behind those of male-led new ventures. Women-led companies tend to be less profitable, with a lower four-year survival rate. She indicated that this may be associated with different motivations between men and women for starting their businesses. She suggested that women start businesses to achieve more job flexibility and autonomy, while men start businesses to achieve economic and financial objectives. Not surprisingly, she further noted that women also face more competing domestic demands on their time. As a result, female entrepreneurs tend to have fewer hours on average to invest in growing their ventures. Thus, Windham-Bannister suggested, these gender dependent preferences and behaviors impact the success and profitability of business ventures.

---

[5] Hazen, Robert M. "Why Should You Be Scientifically Literate?" *BioScience*. 12 (2); Katz, S.J. (2008) "WEBS: Practicing Faculty Mentorship." *Bioscience*. 58 (1): 15.; DeWelde, Kristine, Sandra Larsen, and Heather Thiry. *Women in Science, Technology, Engineering and Math (STEM)*.

She further emphasized that capital funds acquisition can be a daunting process for new businesses, but failure is part of the learning curve. In this vein, she suggested serial entrepreneurs appear to be viewed favorably by venture capital firms and company failure may be considered a mark of having been tested. She indicated this serial behavior is more commonly associated with men and their acquisition of venture capital funding. Interestingly, Windham-Bannister noted that fewer than 10 percent of all requests come from women, leading to women-owned businesses receiving only 3-5 percent of overall venture capital funding. However, she emphasized that 13 percent of female-led firms and 15 percent of male-led firms requesting venture capital funding receive it, thus indicating the disappearance of a gender bias. As a consequence, she stressed that women need to ask for money more frequently. Similarly, she discussed a study of institutions affiliated with the Harvard Medical School that found that their female and male investigators had equal success at winning grants from NIH. However, women requested fewer dollars for their grants ($115,000 on average versus $150,000), and also received fewer dollars ($98,000 vs. $120,000). In both science and business, she suggested that women need to ask for more.

Windham-Bannister additionally commented on the perspective of investors. She indicated that the amount of funding an entrepreneur requests from funding sources is often a critical indicator of industry knowledge. Specifically, she noted that the monetary amount may indicate whether or not the start-up firm truly understands what it will take to start and grow a successful business. So, she suggested that it might be easier asking for more money rather than less; entrepreneurs cannot afford to sell themselves short when it comes to asking for money.

"Investors invest in people," she said. "They would rather invest in a good person who has a lousy idea than invest in a great technology that is being stewarded by someone whom they consider less aggressive, less capable, and less talented." As a result, she suggested that bridging the gap between science and business requires acknowledging the differences in the mindset of science versus the performance metrics of business. In science, the focus is on methodology, validating findings, and openly sharing published results. The measure of success in science tends to be one of prestige and contributions to the body of scientific knowledge (e.g., publications). In business, the focus is on results and winning in the marketplace, with value equating to a competitive advantage. This places demand on the utility of a new technology or product. Success is measured in financial performance and shareholder value.

Windham-Bannister suggested that for those scientists who aspire to be entrepreneurs, there is a need to blend these paradigms by understanding the demands of both science and business. She emphasized that most who have come up through the science, technology, engineering or mathematics paths have gone through a training process that was anchored in data and analysis, was highly respectful of critical thinking, and favored deductive reasoning and validation. The world of business demands "emotional intelligence," meaning that an entrepreneur understands not only their own emotions, but also those of their stakeholders. She urged aspiring scientific entrepreneurs to be willing to leap across the chasm and develop an appetite for risk taking in order to be competitive in the field.

By elaborating on emotional intelligence based on her personal experiences, Windham-Bannister noted that entrepreneurs own the vision and the strategy of a business, which requires intervening quickly to address problems that arise and making hard, fast decisions. She suggested that these are not skills that scientists, in general, possess. Women in particular may struggle with making decisions without collaboration and seeking outside opinions. She underscored that women trained in science, technology, engineering and mathematics are comfortable with details, data, and lots of information, but as an entrepreneurs, the focus needs to shift to the big picture. Therefore, to aid in changing this mindset, she stressed the importance of networks and mentors.

Finally, Windham-Bannister addressed the problem that many women entrepreneurs face of having limited networks. She suggested that they utilize available male mentors and their networks, as well as those associated with women. She went on to discuss gender differences in professional networks and mentoring. Specifically, she discussed a study that showed that Caucasian men receive more information about opportunities of all types—jobs, money, networking opportunities—through their networks than do women or racial minorities. To cover such information gaps, she encouraged women to find a support-system not only to gain a strong network, but also to learn from their mentors' skills and experience.

3

PANEL II: ASPECTS OF LEADERSHIP IN BIOTECHNOLOGY CAREERS

Judy Heyboer, Human Resources Consultant, Former Senior Vice President, Genentech, Inc.
Barbara Wallner, President and CEO, Chymic Therapeutics, Inc.
Lydia Villa-Komaroff, Chief Scientific Officer, Cytonome/ST, LLC

Judy Heyboer, Human Resources Consultant, Former Senior Vice President, Genentech, Inc.

Judy Heyboer focused her address on the key attributes in business. She prefaced her presentation by noting that her comments were derived from information acquired through conversations within her personal network of professionals that includes CEOs, executive coaches, and human resources professionals. Heyboer underscored that to be successful one needs to portray confidence, competence, and potential. She further concluded that successful entrepreneurs exhibit a number of key behaviors and attributes. These include:

> "Assume equality...it is essential to success."
>
> -Judy Heyboer, Human Resources Consultant and Senior Vice President (retired), Genentech, Inc.

- contributing effectively and succeeding in a team environment
- understanding the working culture of an organization and adapting accordingly
- being open to and inclusive regarding organizational politics
- using a collaborative approach to projects, but staying accountable for personal roles in these efforts
- taking initiative and risks rather than waiting until tasks are delegated
- asking questions and listening to the answers
- being willing to make mistakes and correct one's course
- getting in a mindset that begins with "yes" instead of "no"
- effectively delegating tasks, aligning tasks to your goals, and structuring an approachable follow-up procedure
- utilizing role models, mentors, and networks to their fullest
- taking responsibility for personal growth and advancement

Heyboer further discussed the challenges faced by women in leadership positions. She emphasized that a number of traits and behaviors are assets to female leaders, but can also be detrimental. For example, Heyboer noted that empathizing with individuals is a positive trait; however, occasionally women will over-tolerate, wait too long, and allow individuals to continue too long before they address an issue. Similar behavior such as self-disclosing, being willing to admit faults, being organized, and wanting to be liked should all be monitored to prevent assets from turning into liabilities. Heyboer encouraged women entrepreneurs to assume gender equity exists, to assume that the responsibility for removing themselves from bad situations, to retain a sense of humor, and to remain positive.

Barbara Wallner, President and CEO, Chymic Therapeutics, Inc.

During her 30 years in the biotechnology industry, Barbara Wallner personally observed dramatic increases in the confidence of women within the field, and in their ability to begin independent careers. Wallner noted that as a successful women leader, a few key personal traits have aided her, including passion, resilience, flexibility, diverse experiences, and extensive networks.

Within the context of women entrepreneurs, Wallner also drew on her personal experience to discuss the critical aspects of successful biotechnology companies. She first outlined six key attributes of a successful biotech company as seen in Box 3-1.

Box 3-1: Six Key Attributes of a Successful Biotech Company

1. Strong, marketable, and well-tested technology with a high potential of success.
2. An enthusiastic and experienced management team, and a strong advisory board.
3. Scientific and medical advisors to better assess potential markets.
4. A strong board of directors that will influence stakeholders' decisions to fund the venture.
5. Strong and reliable investors who trust the venture and will provide funds regularly.
6. Motivated and well-trained employees who have a stake in the success of the company.

Wallner next explained the importance of a strong business plan that depends on: the strength of the founding technology, a clear and accessible market of the relevant size, a favorable financial environment, investor confidence, and a defined exit strategy.

She discussed three personal business case studies—Biogen, Point Therapeutics, and Chymic Therapeutics—where scientific research was successfully converted into business ventures. Specifically, Wallner noted that through her experiences with these companies, she was able to learn how to take science into business and to assess whether a certain aspect of science has a commercial value. She noted that her first job with Biogen was ideal because the atmosphere was similar to that of a research institution, where she and her colleagues were free to conduct research and to come up with new programs. This allowed her to move across scientific fields, which helped her understand that one does not need to stay within one's own scientific discipline. The ability to move across fields is essential to running programs and evaluating what is good for companies.

Wallner shared a concrete example of how she developed and exhibited leadership skills: In order to develop a technology for a company, she personally had to recruit the scientists who would work on the project by going to each one and encouraging them to become involved. She noted, however, that not all products and technologies worth bringing to market actually reach that goal. She has experienced situations when the products were great, and the clinical trials were great, but the markets changed and/or the funding realities changed. Therefore, she noted the importance of remaining passionate, and seeking alternate sources of funding because often "other opportunities are right around the corner."

Within each of these examples, Wallner emphasized the key leadership traits discussed above that aided her in these entrepreneurial endeavors. In her opinion, passion, resilience, having an open mind, and the ability to think outside-the-box have all aided her in developing a successful career.

Lydia Villa-Komaroff, Chief Scientific Officer, Cytonome/ST, LLC

Lydia Villa-Komaroff began by questioning the idea of how success is defined, and referenced *Outliers: The Story of Success* by Malcolm Gladwell as an inspiration for her address.[1] Specifically, she suggested that success is not a consequence of individual merit, but rather results from cultural legacies, hidden advantages, and extraordinary opportunities. Villa-Komaroff reviewed her own career path, and presented some cumulative advantages she experienced that have led to her personal success. Specifically, she noted that as a Mexican-American and eldest child, she learned about competition and collaboration at an early age. Villa-Komaroff further emphasized the timing of her doctoral research and the emergence of her field, especially in Boston, that led to unique opportunities. She further underscored the role of mentors at every stage of her career that helped to push her forward along her path. She also personally undertook innovative business strategies to navigate through challenging real-life situations that arose in the companies with which she was involved, some of them were considered unconventional at the time.

Villa-Komaroff noted the importance of persistence and hard work to success. Citing Gladwell, she noted that it takes approximately 10,000 hours of practice for someone to master a field. In addition to this practice and hard work by individuals, there are specific actions that can be taken "to ensure that people of talent can live up to their talent," including adopting policies that can help them succeed, changing our own behavior, and undertaking formal activities such as mentoring, networking, and training.

---

[1] Gladwell, Malcolm (2008). *Outliers: The Story of Success*. New York: Little, Brown and Company.

4

PANEL III: EDUCATION TO PREPARE FOR ENTREPRENEURIAL CAREERS

Sheldon M. Schuster, President, Keck Graduate Institute of Applied Sciences
Gail Naughton, Dean, College of Business, San Diego State University and Founder, Advance Tissue Science, Inc.
Michael Teitelbaum, Program Director, Alfred P. Sloan Foundation
Jessica Townsend, Assistant Professor of Mechanical Engineering, Olin College

Sheldon M. Schuster, President, Keck Graduate Institute of Applied Life Sciences

Sheldon M. Schuster began by discussing the need for management-oriented professionals in the field of life sciences, based on discussions with life science industry representatives. The Keck Graduate Institute of Applied Life Sciences (KGI) was founded in response to an industry-driven need to bridge the gap between research and management. Schuster emphasized that science does not explicitly drive the life science/biotechnology sector, rather business does. Schuster presented the following framework, shown in Figure 4-1, which reflects the role of risk in determining the trajectory of a business. Specifically, he noted that the market drives industry, so adaptability is paramount and success is defined as moving the market forward.

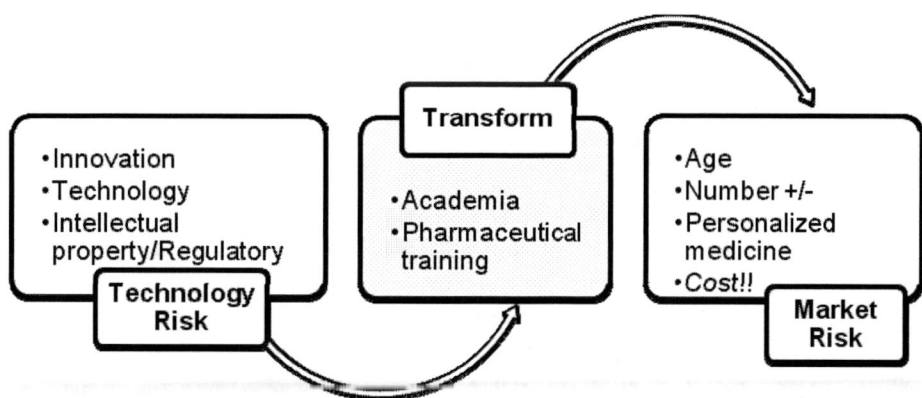

FIGURE 4-1 Industry diversity framework.
SOURCE: Slide presented by Sheldon Schuster at *From Science to Business: Preparing Female Scientists and Engineers for Successful Transition into Entrepreneurship Workshop* on August 31, 2009.

Schuster further noted the gap in perception from academia to industry. He explained that in a recent survey, 92 percent of the postdoctoral life science fellows believed that their role in industry would be conducting basic science research. However, Schuster quickly added that in reality only 8 percent of jobs in the life science industry were in this area. Instead most jobs exist in other areas such as regulatory affairs, project management, corporate communications, quality assurance, clinical trials, production, intellectual property, corporate development, and marketing. Schuster explained that to perform in areas outside of basic science research, a technical background is necessary in addition to a thorough understanding of business. He went on to note that recent doctoral graduate students or postdoctoral fellows typically develop broad scientific knowledge and quantitative reasoning, but often do not develop a variety of other skills that industrial representations desire: ethical judgment, intercultural skills, teamwork, critical thinking, and adaptability.

According to Schuster, for researchers who have received a life sciences degree in the last ten years, most are currently postdoctoral scholars and earn roughly $38,000 a year. Schuster suggested that the current system is failing and doctoral students need to be educated differently. Schuster discussed the Postdoctoral Professional Masters in Bioscience Management program at KGI. He described the curriculum as a broad, team-oriented, project-based program, which focuses on enhancing the leadership and professional development skills of postdoctoral scholars, and providing extensive co-curricular activities. Schuster emphasized that the program is geared toward individuals who want to change their position and make an investment in their future.

Gail Naughton, Dean, College of Business, San Diego State University and Founder, Advance Tissue Science, Inc.

Gail Naughton focused on educational programs that strive to better prepare scientists for roles in biotechnology and high-technology fields. She first provided a background context for the biotechnology field by emphasizing that scientists need to understand business. Further, she continued that business managers need to understand both scientific and regulatory components in order to be successful. As a consequence, Naughton stressed that educators need to teach students in the biotechnology field all of the parameters that contribute to a successful business venture, so that they are adequately prepared to overcome the high failure rate in the market.

To cater to this need, Naughton discussed efforts at San Diego State University, where they have launched a variety of degree programs, including an MBA in biotechnology and life sciences and a joint Ph.D./MBA degree program. She emphasized that one of the challenges with these programs is finding biotechnology-specific case studies. Negative data is rarely published, with the result that biotechnology entrepreneurs continue to make the same mistakes over and over again. Naughton stressed that in order to instill the passion for life science entrepreneurship and to train the next generation with the tools to be successful, it is important to write about the risks of these ventures, even in instances of limited triumph. Additionally, these business-driven degree programs highlight the role of intellectual property, discuss bio-ethics, place active venture capitalists in the classrooms to directly interact with students, and include finance courses focused on challenges related to the biotechnology field. Naughton further elaborated on the use of on-line courses to create flexible educational experiences for currently employed professionals in fields such as quality control and regulatory affairs.

She further underscored the proximity of these programs to the city of San Diego, where the thriving biotechnology industry enables students to interact closely with local professionals in order to build their personal networks. Moreover, Naughton noted that the biotechnology industry is becoming increasingly global. In response to this trend, the global entrepreneurship MBA at San Diego State University allows students to travel to China, India, and the Middle East during their educational experience. Naughton commented that this adds an additional layer to the entrepreneurial educational experience, as students interact with corporate partners and national laboratories to understand risk-taking on a global scale. She commented that all of the programs she discussed stress an entrepreneurial mentality that recognizes that employees should not be limited to specific tasks, but rather need to be able to adapt to any task that needs completing. This requires that each employee have a basic understanding of the entire company's structure and practice.

In closing, Naughton emphasized the need for entrepreneurship-related education. Specifically, she stressed the need for biotechnology-specific business case studies and greater student interactions with industry leaders to develop networks. Naughton recognized the importance of increasing internship opportunities, of engaging students in this field at a younger age, and preparing students with the foundation to explore their passion in search of related opportunities.

> "Sometimes ventures work, sometimes they don't, but they are always very valuable learning experiences."
>
> -Gail Naughton, Dean of the College of Business, San Diego State University and Founder, Advance Tissue Science

Michael Teitelbaum, Program Director, Alfred P. Sloan Foundation

Michael Teitelbaum focused his talk on Professional Science Master's (PSM) programs that are "designed to prepare students for entrepreneurial careers." At the time of this presentation, 77 participating PSM degree-granting universities exist in 25 states, with approximately 2,600 enrolled students and 2,600 alumni. Participation and program development has recently accelerated in certain regions and states, including Arizona, California, Florida, Illinois, New York, New Jersey, and Oregon. Teitelbaum discussed the Sloan Foundation's involvement in promoting such educational opportunities, as one of the few foundations devoted to science, especially at the advanced research level. He explained that the Sloan Foundation has expanded its programming efforts beyond academic career preparation by encouraging industry-oriented programs such as the PSM.

Teitelbaum indicated that the Sloan Foundation's emphasis on PSM degrees is a result of the recent shift away from a sole focus on research-based careers among science and math students. He noted that approximately 80 percent of graduating doctoral students with science and math majors choose careers away from academia. He continued that given the foregone income, career risks involved in doctoral studies, and disrupted work-life balance, students - especially United States citizens - increasingly find doctoral degrees less attractive. Science-intensive employers, however, continue to demand workers with graduate-level science backgrounds as well as skills in project management, interdisciplinary work, computation, legal and regulatory fields, ethics, and a basic business background. Thus, pure research-focused doctoral programs are often inadequate. The importance of a PSM degree is underscored by the fact that both corporate and government sectors are facing economic challenges, which lead to rapid product shifts and require businesses and their employees to be flexible and nimble. Additionally, globalization and off-shoring have created new challenges regarding long-term project trajectories and competition in low-cost emerging markets.

Teitelbaum further discussed the national status of the PSM degree to date. There are 145 U.S. PSM programs, indicating an increase in the number of programs and students. With regard to gender, Teitelbaum noted that a higher number of women are attending PSM programs than other programs: 47 percent of students in 2008 were women versus 38 percent who sought doctoral degrees and 40 percent that sought master's degrees in natural sciences.[1] The PSM degrees awarded in 2008 were further broken-down by discipline and disaggregated by gender to illustrate gender equity in almost all disciplines, except for financial mathematics and applied physics, as shown in Table 4-1.

TABLE 4-1 Percent of Professional Masters Degrees Awarded, by Gender and Discipline, 2008

Discipline	Male	Female
Bioinformatics/Biotechnology	54%	46%
Microbiology/Cell and Molecular Biology	45%	55%
Financial Math/Industrial Mathematics	63%	37%
Applied Statistics/Computational Science	58%	42%
Analytical Chemistry/Biochemistry/Forensics	50%	50%
Food Safety/Pharmacology and Toxicology	50%	50%
Environmental and Geosciences	50%	50%
Applied/Industrial Physics	100%	0%
Health/Medical Physics	42%	58%
Total PSM Average	55%	45%

SOURCE: National Science Foundation, Division of Science Resources Statistics, 2008. *Science and Engineering Degrees: 1966-2006.* Detailed Statistical Tables NSF 08-321. Available at http://www.nsf.gov/statistics/nsf08321/.

Teitelbaum hypothesized a possible rationale for the small gender gap based on interviews of current and former PSM students, which revealed that women were attracted to the program because of its fixed, compact nature and a perceived better work-life balance compared

---

[1] National Science Foundation. (2008) *Science and Engineering Degrees: 1966–2006.* Detailed Statistical Tables NSF 08-321. **NPSMA, 2008.

to doctoral programs. It further provided a pathway for career re-entry after prolonged time away from the field. The students indicated that the skills developed during the PSM degree were versatile and portable, thus encouraging career flexibility. Finally, they emphasized that the opportunity for internships provided real-world work experience that enabled them to negotiate career transitions. In this way, PSM programs are able to encourage science-related entrepreneurship by attracting students who may otherwise choose not to pursue these careers.

Finally, Teitelbaum stated that PSM degrees have received tremendous support and enthusiasm from industry in the form of funding, placements, and mentorship opportunities. He noted that government agencies have launched equally encouraging programs through the American Competes Act and the National Science Foundation. Through this continued support, Teitelbaum suggested that the expectations for PSM degrees remain high and the Sloan Foundation envisions the PSM degree becoming a "normal" graduate degree and a favorable pathway for science entrepreneurs.

Jessica Townsend, Assistant Professor, Mechanical Engineering, Olin College

Jessica Townsend introduced the unique academic and engineering training space created by the 1997 charter at Olin College. She explained that Olin College has specifically addressed the undergraduate-level educational challenges associated with merging science, engineering and business. As a result, all students at Olin College are required to take three classes with an entrepreneurship focus:

- Fundamentals of Business and Entrepreneurship
- User Oriented Collaborative Design
- Senior Capstone Project for Engineering (SCOPE)

All first-year students complete the Fundamentals of Business and Entrepreneurship, where students interact directly with successful entrepreneurs, learn business basics, and are expected to use their advice to complete a project that challenges students to develop a new small venture. The second course, User Oriented Collaborative Design, is focused on identifying needs and opportunities within a group of people. For example, one group of students completed a course project about the needs of the bike messenger profession. The students spoke with bike messengers, learned their values, and even attempted bike messengering. Later, the students brainstormed products that could be developed to meet the needs of bike messengers and pitched them to bike messengers for feedback. Townsend emphasized that this course immersed the students in the fundamentals of understanding a market and determining technologies that would be relevant and successful. In the third experience, SCOPE, businesses challenge groups of approximately five students who are asked to tackle a real-world industry problem over the course of an academic year. Similar to above, students determine the values of their customers and then generate products to meet the determined needs.

Townsend further explained that students have the option of specifically concentrating on entrepreneurship. Students

> "When women have an idea, they want to share. When men have an idea, they want to start a business."
>
> -Jessica Townsend, Ph.D., Assistant Professor of Mechanical Engineering, Olin College

who choose this path take additional elective courses and complete a second SCOPE course that requires them to generate a business plan either alone or in a team. Interestingly, at this stage a gender divide is observed: while Olin faculty and student gender ratios approach equity, with women constituting 46 percent of the student body, just 25 percent of all students choose to add an entrepreneurship concentration and only 25 percent of those in the concentrating program are women.

Townsend suggested that this gender gap may result from the structure of the Olin concentration program, where students choose either an entrepreneurship or an arts/humanities concentration. Therefore, the numbers may not reflect a lack of women's support or interest in entrepreneurship, but rather just be a factor of a difficult choice. Townsend remarked that compared to men, women are less likely to convert their business ideas into new ventures and often find it difficult to be assertive at networking events. Finally, Townsend stressed that her female students continually comment on the importance of hearing from female guest-speakers and other female role models as imperative to their confidence in pursuing entrepreneur paths.

5

STUDIES ON ENTREPRENEURSHIP

*Caroline Simard, Director of Research and Executive Programs,
Anita Borg Institute for Women and Technology*
Manwai ("Candy") Ku, Researcher, Stanford University

Caroline Simard, Director of Research and Executive Programs, Anita Borg Institute for Women and Technology

Caroline Simard discussed an Anita Borg Institute and Clayman Institute study, *Climbing the Technical Ladder*,[1] which looked at technical women in Silicon Valley high-tech companies. Population samples of 800 male and female employees at seven Silicon Valley companies from all technical levels were surveyed to understand the advancement of women from the entry level to senior leadership roles. Survey results indicated that across all levels, women constituted 20 percent of the technical employees, mostly software and hardware engineers, and were not equally represented at all technical levels. Instead, significantly more women were likely to be entry level, compared to only 4 percent of women reaching leadership roles in the companies surveyed. Furthermore, technical women were more racially and ethnically diverse than technical men, shown in Figure 5-1.

However, Simard commented on a number of areas where gender equity existed. The average age of individual company starters is approximately 41 and these individuals have roughly 15 years of previous work experience. Simard further noted that the educational profile of the male and female employees was similar, suggesting that the poor advancement of women is not due to educational barriers.

---

[1] Simard, C. et al (2008). *Climbing the Technical Ladder: Obstacles and Solutions for Mid-Level Women in Technology*. Retrived from http://anitaborg.org/files/Climbing_the_Technical_Ladder_Exec_Summary.pdf, March 9, 2012.

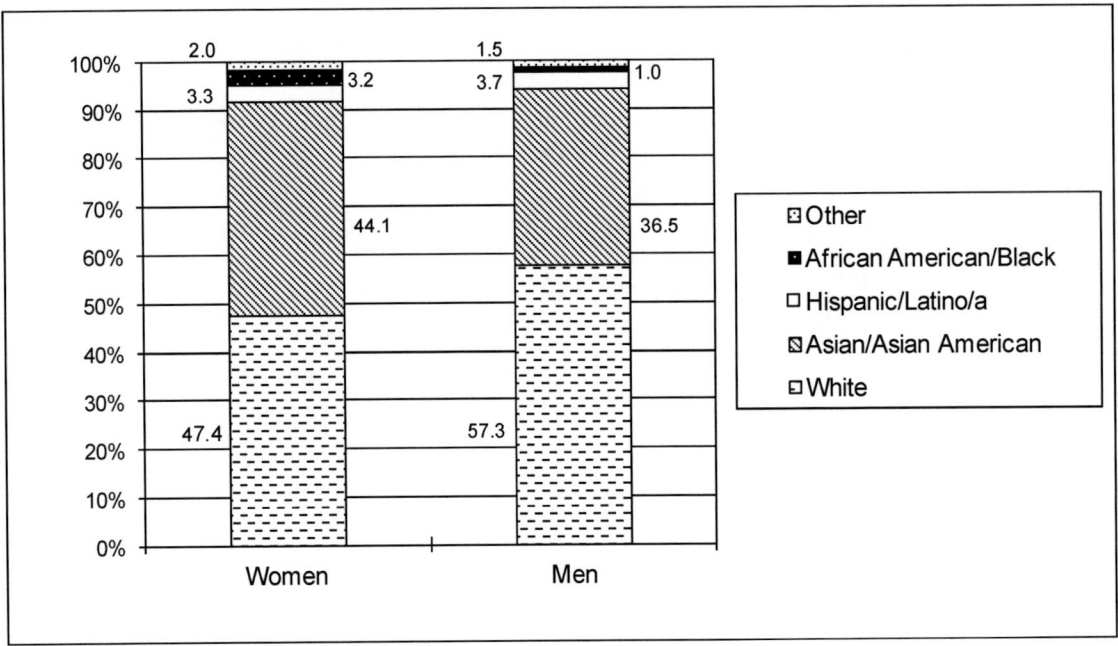

FIGURE 5-1 Race and ethnicity of technical workforce, by gender.
SOURCE: Simard, C. et al (2008). *Climbing the Technical Ladder: Obstacles and Solutions for Mid-Level Women in Technology.* Copyright granted by Anita Borg Institute for Women and Technology.

To describe the factors that lead to success in high-technology companies, Simard discussed survey results that probed employee perceptions about themselves and their coworkers. Men and women noted that in order to be successful, one must be analytical, innovative, questioning, risk-taking, collaborative, entrepreneurial, and assertive. Simard emphasized that the combination of these attributes suggests that a specific, assertive communication style may be preferred in order to advance in high-tech firms. In terms of personal perception, both men and women perceived themselves similarly as analytical, assertive, and risk-taking. However, significant gender differences were observed when determining self-perceptions of being innovative, entrepreneurial, and collaborative.[2]

Furthermore, women and men reported considerable differences in their belief that successful mid-level individuals need to work "long hours" in order to be successful, as shown in Figure 5-2. She stated that more women than men believed this attribute to be necessary for individual success, but fewer women perceived themselves to be working the many hours they determined necessary in order to be successful, while more men believed that they work as many hours as required for success.[3]

---

[2] According to a 2010 report by the Anita Borg Institute, *Senior Technical Women: A Profile of Success,* only 29.6 percent entry/mid-level women describe themselves as an "innovator," versus 38.1 percent of senior women and 60.2 percent of senior men. In addition, less than half of high-level technical employees in large companies perceive themselves as entrepreneurial (31.7 percent of women versus 40.5 percent of men).

[3] *Senior Technical Women: A Profile of Success* indicated that senior women were significantly more likely than women at the entry and mid-levels to perceive themselves as working long hours.

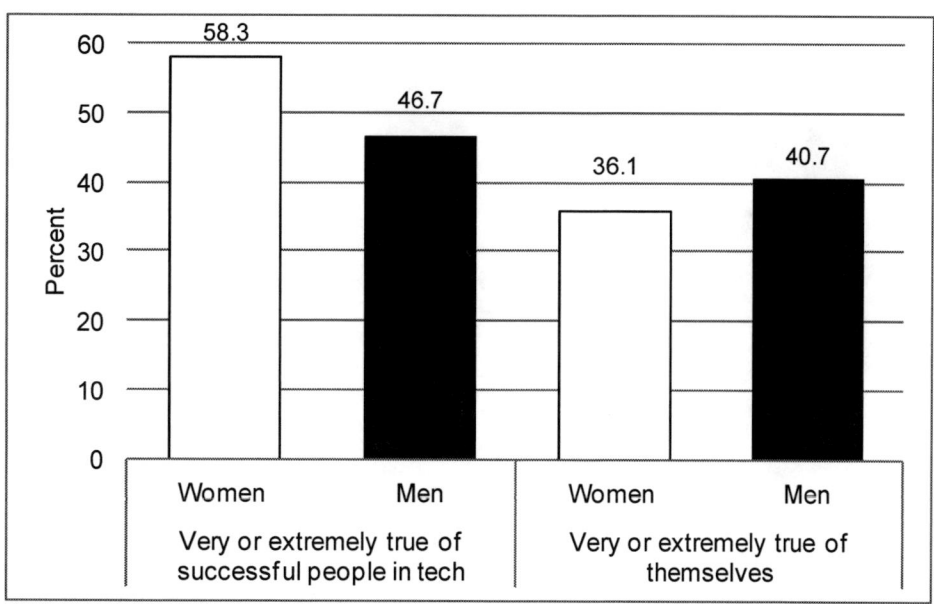

FIGURE 5-2 "Long working hours" Attribute: Attribute of success versus self-perception among mid-level technical men and women.
SOURCE: Simard, C. et al (2008). *Climbing the Technical Ladder: Obstacles and Solutions for Mid-Level Women in Technology*. Copyright granted by Anita Borg Institute for Women and Technology.

Simard next discussed employee feedback about their managers. In almost all attribute categories, no difference between male and female managers was observed except for their perceived technical skills. Both male and female employees rated their female managers lower in technical skills than their male manager counterparts. Simard explained that both men and women tend to exhibit an implicit bias against the technical skills of women. She further emphasized that this is increasingly important, considering that high-tech companies view technical skills as a critical indicator of success and, therefore, a key point for intervention to advance women.

In looking at family configurations, Simard noted that technical men and women appear to have children at similar rates, but gender differences arise when looking at the overall family structure. As shown in Figure 5-3, 80 percent of technical women, but only 40 percent of technical men, have a partner who also works full-time. This industry profile differs significantly from typical U.S. households, where only 19 percent of all marriages are based on the woman staying home and the man working. Simard stated that these results suggest that the primary household responsibilities tend to fall predominately to women in technical sector families leading to an unequal distribution of family responsibilities. She noted that this gender difference in family configuration may explain why women frequently do not seek upward mobility. Simard suggested that typical advancement rewards are delegated to individuals whom are available at all hours, so with increased household responsibilities, technical mid-level women do not have the same freedom to allocate their time as their male counterparts. She further noted that this trending is similar across all levels. In addition, most women are also partnered in dual-technical households.

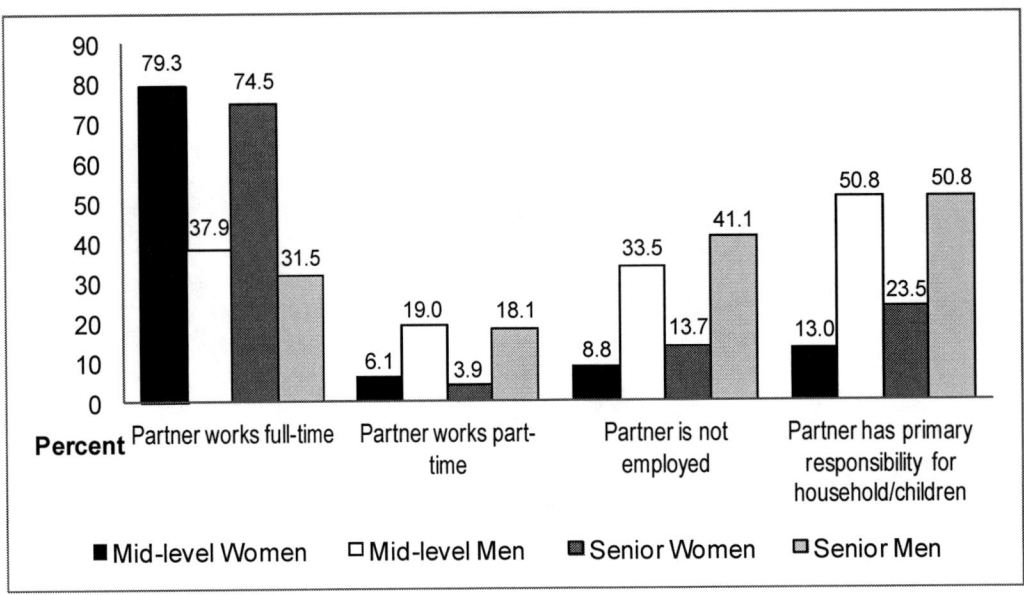

FIGURE 5-3 Household characteristics of partnered mid-level and senior-level technical workers, by gender.
SOURCE: Simard, C. et al (2008). *Climbing the Technical Ladder: Obstacles and Solutions for Mid-Level Women in Technology.* Copyright granted by Anita Borg Institute for Women and Technology.

After understanding these gender differences, Simard discussed the future plans of these employees. She found that men and women have similar aspirations in the next 12 months, except men were more likely to indicate they planned to start their own company. Although the absolute numbers were small for both men and women, less than 10 percent, a gender divide was observed. She further noted that men and women who ranked themselves higher regarding the attributes of long working long hours, being innovative, and being entrepreneurial were more likely to indicate that they intended to start a company in the coming months. Interestingly, these were the three attributes about which women overall ranked themselves lower than men. She concluded that there are technical women who are willing to take risks, but do not view themselves as innovators or entrepreneurs. As a consequence, interventions with respect to these self-perceptions, as well as providing adequate support mechanisms for dual family careers, are necessary to close the existing gender gaps.

Manwai ("Candy") Ku, Researcher, Stanford University

Manwai (Candy) Ku discussed the findings of her research focused on entrepreneurship and gender gaps in the high-tech industry.[4] Ku tracked venture-backed technology companies in

---

[4] More recent research by Ku is available in Justine Tinkler, Manwai C. Ku, Kjersten Bunker Whittington, and Andrea Davies. (Forthcoming) "Gendered Decision-Making: Assumptions about Technical Knowledge and Social

Silicon Valley and found women to be underrepresented in various ways. In 2006, women CEOs led only 4 percent of information technology related business ventures and, as of 2007, women owned less than 5 percent of information technology firms in Silicon Valley. Ku suggested that both human capital and social capital barriers contributed to these gender differences in securing the venture capital that leads to increased entrepreneurship. Specifically, fewer women enter the technology sector and they participate in smaller venture capital networks. Although increasing the number of women in the technical pipeline and increasing their networking options may be solutions to closing the gender gap, Ku suggested that cultural assumptions and double standards may complicate these efforts. Some venture capitalists may question the technical competence of women, which causes them to experience a double standard. Women believe they need to do more as well as make fewer mistakes in order to seem as competent as male counterparts.

Ku and colleagues probed these biases through a social-psychological experiment, in which a real-world business plan was translated into four versions, changing only the biography of the entrepreneur who would be founding the company in two ways; gender and technical experience were varied (male versus female, and computer science versus history backgrounds). All entrepreneurs were described in the business plans as possessing an MBA and six years of relevant experience and were said to have previously been a vice president of a start-up company. Members of the Stanford Business School Entrepreneur Club acted as venture capitalists, assessed the project proposals, and were told that if their results matched those of actual venture capital firms, they would receive greater monetary rewards. The specific aspects of the proposals that club members were asked to evaluate were: the venture, the entrepreneur, and the influence of the founder's contacts.

Ku found that the gender and technical background of the applicants significantly affected the final decision of the club members. In the evaluations of the venture, the existence of technical backgrounds was influential: products proposed by entrepreneurs with technical backgrounds were rated higher than the products proposed by entrepreneurs with non-technical backgrounds. The presence of a technical background was shown to aid both men and women in the evaluations of the venture with reviewers more likely to request additional meetings, buy the product, or invest in the venture offered by those with technical backgrounds as opposed to those without. Ku suggested that this supports the classic human capital component; an increase in the number of technical women may lead to an increase in the number of female entrepreneurs.

A clear trend was not evident in the evaluations of the entrepreneur. While having a technical background helps both men and women overall, interestingly, when assessing evaluations of the entrepreneurs themselves the picture is more complex. Having a non-technical background caused women to be rated lower than women with technical backgrounds, however, men without a technical background appeared to be rated higher than women without a technical background perhaps because they were perceived to be more savvy. These findings suggest a possible double-standard in the evaluation of human capital.

A further interesting finding was that when women evaluated the entrepreneur, they rated the non-technical women the highest and best able to penetrate the market. Ku suggests that these results support the hypothesis that venture capital firms with female partners are more likely to invest in women-led ventures than firms without female partners. Similarly, the

---

Capital in Venture Capital Evaluations." *Gender and Society*. Submitted for publication at time of this publication. Working paper available upon request.

influence of the founder's contacts provided greater benefit to women than men, implying a double standard with regard to social capital as well.

Ku noted that these results show evidence of subtle assumptions about gender as well as support for the classic human and social capital arguments. She suggested that such gender discrepancies may be overcome by focusing on both supply-side and demand-side factors. Ku suggested that increasing the representation of women in technical fields and promoting networking opportunities for women may help supply-side factors. She further suggested that demand-side biases may be overcome by promoting gender awareness among venture capitalists and by encouraging more women to enter the venture capital industry.

6

PANEL IV: ALTERNATIVE FORMS OF ENTREPRENEURSHIPS IN SUSTAINABLE TECHNOLOGIES: INTRAPRENEURSHIP IN CORPORATIONS AND GOVERNMENT, SOCIAL ENTREPRENEURSHIP, AND TRADITIONAL ENTREPRENEURSHIP

Sharon Nunes, Vice President, IBM Green Innovations
Maxine L. Savitz, General Manager for Technology Partnerships, Honeywell Inc., Retired, and former Deputy Assistant Secretary for Conservation, U.S. Department of Energy
Judith Giordan, Senior Advisor, National Collegiate Innovators and Inventors Alliance
Lucinda Sanders, CEO and Co-founder, National Center for Women & Information Technology

Sharon Nunes, Vice President, IBM Green Innovations

Sharon Nunes addressed the occurrence of entrepreneurship within a large company, that she termed *intrapreneurship*. Specifically, she commented on the factors that contribute to success in this arena and emphasized the need for understanding the markets to help identify future gaps, trends, needs, and technical developments. She commented that it is exceedingly important to be able to articulate a vision to effectively "sell" the idea, to know how the project will make money, and above all, to deliver. She noted the importance of diversifying the project team to include those with technical, financial, and operational skills, and market analyzers, and other professionals who complement one another in order to create a successful leadership team. She indicated that dedication, perseverance, confidence, passion and an appetite for risk were other necessary attributes for successful intrapreneurship.

Nunes stated that significant challenges exist for the concept of intrapreneurship, beginning with the notion that innovation only occurs in small, nimble companies. Large companies may provide a larger safety net for ventures and steady employment that may appeal to a greater number of people. Therefore, she emphasized the importance of risk-taking, but cautioned against avoiding market signals.

Finally, to encourage women's participation in the field, she offered a few suggestions. She remarked that women who began their careers at large companies, and who have advanced to leadership roles, may increase the participation of women in venture capital endeavors. She commented that women are particularly interested in social entrepreneurship and this may provide a gate-way to increasing female participation in general entrepreneurship and intrapreneurship. Lastly, she commented that persistence is imperative and suggested that once an individual develops a reputation for delivering on their promises, gender differences no longer exist.

Maxine Savitz, General Manager for Technology Partnerships, Honeywell (retired) and former Deputy Assistant Secretary for Conservation at the U.S. Dept. of Energy

Maxine L. Savitz discussed her diverse experiences as an entrepreneur, both in the industry and the government. She specifically elaborated on the entrepreneurial opportunities offered by government agencies at all organizational levels. To help identify the possibilities for creativity and innovation in the government, Savitz drew parallels between industry and the government, highlighting the following features:

- Funding: Congress, like venture capitalists, expects justification of allocated funds. Thus, as in the corporate sector, government workers need to build sound business plans and carry out extensive expenditure analysis before committing their limited resources in a new venture.
- Personality: Both sectors require entrepreneurs to be on the lookout for possible opportunities, be willing to take risks, and be flexible in making the most of an opportunity.
- Collaboration: Linking research, development, and policy-making helps create more efficient products for both private and public sectors. Creative and technical analysis is important for building innovative and efficient public policies that can be readily implemented in conjunction with the needs of industry.
- Trend Identification: As in commercial markets, Savitz suggested that opportunities for government agencies can also be very dynamic. For example, the energy sector has moved from being a mainly private sector commodity in the 1960s, to a majority public sector today. In recent years, the share of industry in utilities is again increasing, necessitating a further shift in government policies.

Finally, Savitz emphasized the importance of passion in identifying those areas of entrepreneurship and intrapreneurship that will lead to one's success.

Judith Giordan, Senior Advisor, National Collegiate Innovators and Inventors Alliance

Judith Giordan began by emphasizing the important differences women can make in business. She noted that Fortune 500 companies with the best records of promoting women to senior leadership were 18 to 69 percent more profitable than median companies in their industries; high gender diversity in the management of European companies led to more profitable stock performance; and companies with three or more women on their board outperformed the competition on all measures by at least 40 percent. These statistics have begun to shape global perceptions of the important role of women in business, where countries such as Norway have instituted mandatory quotas for gender diversity.

Explicitly looking at entrepreneurship, Giordan commented that half of all U.S. investment capital comes from women, but only 10 percent of traditional mutual fund managers and only 3 percent of hedge fund managers are women. She emphasized that even the venture capital community acknowledges an implicit bias towards white males, which presents a barrier for both women and men of color. This, she argued, is demonstrated by the fact that only 3 percent of all venture capital funding goes to companies with women CEOs, with 44 percent of those women-led firms located in the healthcare sector.

Giordan further noted that the best predictor of success for entrepreneurial women is confidence; women with greater opportunities for professional networking were also more successful and satisfied with their jobs, as compared to peers with fewer professional networking opportunities. She commented that male and female perceptions may play a large role in advancement and opportunity development. Specifically, male managers perceive more of an even playing field for women to advance than do female managers. Additionally, female managers have more positive perceptions of women's attitudes toward advancement than male managers. Giordan noted that male managers overestimated the amount of home-related stress experienced by women relative to the amount of conflict reported by women themselves. She suggested that this leads to women being passed-over for career opportunities because of perceived time constraints induced by family responsibilities. She urged women to gain confidence, consider themselves as equals, and be proactive in invoking change. Giordan stressed the importance of diversity and its ability to lead organizations to better decisions and better actions. Women and men may differ in their propensity for collaboration, seeking long-term results, competition, and risk-taking, but these attributes are all valuable, and a balance among them is imperative.

Social entrepreneurship, for-profit entrepreneurship, and intrapreneurship all require change. Giordan emphasized that in order to create change, women need to be proactive and take action. She stressed that it is critical that women build both personal and business value to develop entrepreneurial credibility. Giordan emphasized the importance of understanding personal motivations and objectives; service driven ventures are valuable and legitimate, for example, but they do not lead to profits comparable to those produced by venture capital funds for other industries. She noted that for women aiming to succeed in the latter, they need to seek early-stage funding. Less than 10 percent of all angel funding proposals were submitted by women, but were funded at equal rates to men, suggesting that women need to ask more frequently in order to close the gender gap.

Lucinda Sanders, CEO and Co-founder, National Center for Women & Information Technology

Lucinda Sanders reflected briefly on the status of women in information technology (IT) companies, where women start less than 5 percent of all new companies, hold less than 5 percent of all IT patents, and occupy less than 5 percent of all corporate technology leadership positions. Looking forward, Sanders discussed her observations from a series of interviews with women serial entrepreneurs, intrapreneurs, and social entrepreneurs. During the 40 interviews conducted with successful technical women, Sanders noted that the interviewees consistently touched on the following themes.

- They did not recognize themselves as successful in math and science as young girls, but rather evolved into technical innovators at later career stages.
- All cited the importance of mentors, beginning with their parents and recognized mentors who aided in starting ventures, risk-taking, and power-sharing.

- The women indicated that success in entrepreneurship results from the following attributes: patience, passion, persistence, and teamwork.
- Most found work-life balance difficult, but recognized that success was about integrating work priorities into life.
- None of the women mentioned gender issues on their own accord when discussing entrepreneurship, and the interviewers did not initiate discussion of this topic.

Sanders concluded that these female IT innovators evolved over time both in their interest in entrepreneurship and in their confidence to succeed. She then drew on her own experiences, first as an intrapreneur, and currently as a social entrepreneur, to emphasize the differences between such endeavors. Specifically, she commented that as a social entrepreneur, resources are limited and the jobs may be exhausting, but the rewards are high and the passion for making a difference is important. She reiterated that providing educational opportunities may be the logical next step for motivating and preparing the next generation of entrepreneurs, intrapreneurs, and social entrepreneurs.

7

THEMES FROM THE WORKSHOP AND CLOSING REMARKS

In closing, Lydia Villa-Komaroff identified some of the issues raised by various speakers and workshop participants. She began by reflecting that the younger generations have grown up differently than those previous, and therefore one can anticipate changes in the data presented throughout the workshop regarding women's involvement in various entrepreneurial careers and fields as the women who are in their 20s and 30s make decisions as to where to go in their career pathways.

She noted that while a great deal of data was presented and more is always helpful, there are sufficient data upon which to begin to act. Specifically, in order to ensure that individuals of all backgrounds utilize their talents, it may not be enough to show people the data about how it is beneficial to have women actively involved, we may need to mandate change in behavior first and then buy-in will follow. Villa-Komaroff reminded everyone that we also need to be aware of our own biases and, perceptions of ourselves, and how these biases and perceptions can interfere with the ability to use our own talents. Courage, she stated, is not an absence of insecurity or fear, it is action in the face of those insecurities and fears.

Villa-Komaroff concluded by suggesting actions that could follow from the workshop:

- sharing the presentations and data presented at the workshop
- informing our own activities based on what we have learned
- remaining a group that can continue to remind people of the importance of these issues
- developing incentives and disincentives that can change behavior in the short term
- bringing people to the table to demonstrate how the inclusion of women will benefit our entire collective efforts

Members of the audience then offered their personal and professional experiences that enriched this session. Some of the points made by various participants include the following.

- Significant commonalities exist between women in technical fields of academia and entrepreneurship, e.g., the positive impact for women of the presence of females on hiring/tenure committees or funding interview boards, the exclusion from formal and informal networks that women experience; and the lower dollar amounts for grants and funding levels sought by women.
- There is a need to emphasize the value of productive mentors and networking.
- It is important to include careers as entrepreneurs, intrapreneurs, and social entrepreneurs in graduate and postdoctoral education and training.

- Promotion of training programs of shorter length that allow women to gain new skills and experiences on a relatively short timeline that complement more formalized programs.
- Although national data on gender issues do exist, more regional data on female technical entrepreneurial activities may help focus policy-making efforts on this subject.
- It would be valuable to explore effective ways to incorporate venture capitalists into the discussion of gender inequality and women in entrepreneurship.
- The impact of lifestyle and family configuration on gender differences warrants further investigation.
- Effective change can occur by promoting public benchmarking of larger firms and creating policies to tie mentoring to promotional advancement and economic gain.

The final discussion highlighted the wide range of career opportunities available to women in scientific and technical areas beyond traditional academic careers. Through meetings and workshops such as this one, these careers can be better understood and many important lessons can be learned from those who have been successful in entrepreneurial careers. Further, as many of the presenters articulated, careers vary and change over time, so women entrepreneurs should feel free to exit and enter a variety of opportunities as they arise over the span of their professional lives.

APPENDIX A

FROM SCIENCE TO BUSINESS: PREPARING FEMALE SCIENTISTS AND ENGINEERS FOR SUCCESSFUL TRANSITIONS INTO ENTREPRENEURSHIP

Workshop Agenda

August 31 – September 1, 2009
The National Academies
The Beckman Center
100 Academy
Irvine, CA 92617

August 31: Framing Issues and Strategies -- Where We Stand

9:00 am   **Welcome and Introductions**

Lilian Wu, Chair, Committee on Women in Science, Engineering, and Medicine, and Program Executive, Global University Programs, IBM

9:05 am   **Study: Entrepreneurial Careers of Women**

Chair: *Susan Wessler, University of Georgia Foundation Chair, Biological Sciences, University of Georgia*

Speaker: *E. J. Reedy, Manager, Research and Policy, Kauffman Foundation*

9:45 am   **Panel I: From Bench to Business: Career Paths for Ph.D.s**

Chair: *Lydia Villa-Komaroff, Chief Scientific Officer, Cytonome/ST, LLC*

Panelists: *Laurel Smith-Doerr, Associate Professor of Sociology, Boston University; and Lydia Villa-Komaroff, Chief Scientific Officer, Cytonome/ST, LLC*

10:45 am   **Break**

11:00 am	**Keynote**
	Chair: *Vivian Pinn, Director, Office of Research on Women's Health, National Institutes of Health*
	Speaker: *Susan Windham-Bannister, President and CEO, Massachusetts Life Science Center*
12:00 pm	**Lunch**
1:30 pm	**Panel II: Aspects of Leadership in Biotechnology Careers**
	Co-Chairs: *Lydia Villa-Komaroff, Chief Scientific Officer, Cytonome/ST, LLC and Sheldon M. Schuster, President, Keck Graduate Institute of Applied Life Sciences*
	Panelists: *Judy Heyboer, Human Resources Consultant, Former Senior Vice President, Genentech, Inc and Barbara Wallner, President and CEO, Chymic Therapeutics, Inc.*
2:45 pm	**Break**
3:00 pm	**Panel III: Education to Prepare for Entrepreneurial Careers**
	Chair: *Sheldon M. Schuster, President, Keck Graduate Institute of Applied Life Sciences*
	Panelists: *Michael Teitelbaum, Program Director, Alfred P. Sloan Foundation; Gail Naughton, Dean, College of Business, San Diego State University and Founder, Advance Tissue Science, Inc.; and, Jessica Townsend, Assistant Professor of Mechanical Engineering, Olin College*
4:30 pm	**General Discussion and Wrap-Up**
	Allan Fisher, Vice President, Product Strategy & Development, Laureate Higher Education Group
5:30 pm	**Adjournment**

APPENDIX A: WORKSHOP AGENDA

September 1: Moving Forward

9:00 am — **Welcome and Summary of Day 1**

Florence Bonner, Senior Vice President for Research and Compliance, Howard University

9:15 am — **Studies on Entrepreneurship**

Chair: *Pardis Sabeti, Assistant Professor, Systems Biology, Harvard University*

Speakers: *Caroline Simard, Director of Research and Executive Programs, Anita Borg Institute for Women and Technology, and Manwai (Candy) Ku, Researcher, Stanford University*

10:00 am — **Panel IV: Alternative Forms of Entrepreneurships in Sustainable Technologies: Intrapreneurship in Corporations and Government, Social Entrepreneurship, and Traditional Entrepreneurship**

Co-Chairs: *Alice Agogino, Roscoe and Elizabeth Hughes Professor of Mechanical Engineering, University of California, Berkeley, and Allan Fisher, Vice President, Product Strategy & Development, Laureate Higher Education Group*

Panelists: *Sharon Nunes, Vice President, IBM Innovations; Maxine L. Savitz, General Manager for Technology Partnerships, Honeywell Inc. (retired); Judith Giordan, Senior Advisor, National Collegiate Innovators and Inventors Alliance; and, Lucinda Sanders, CEO and Co-founder, National Center for Women & Information Technology*

11:30 am — **Summary of the Conference: Findings and Themes**

Lydia Villa-Komaroff, Chief Scientific Officer, Cytonome/ST, LLC

12:00 pm — **Closing Luncheon**

1:00 pm — **Adjournment**

APPENDIX B

COMMITTEE ON WOMEN IN SCIENCE, ENGINEERING, AND MEDICINE

MEMBER BIOGRAPHIES

Rita R. Colwell (NAS)*, Chair
Rita R. Colwell is Distinguished University Professor both at the University of Maryland at College Park and at Johns Hopkins University Bloomberg School of Public Health, Senior Advisor and Chairman Emeritus, Canon U.S. Life Sciences, Inc., and president and chief executive officer of CosmosID, Inc. Her interests are focused on global infectious diseases, water, and health, and she is currently developing an international network to address emerging infectious diseases and water issues, including safe drinking water for both the developed and developing world. Dr. Colwell served as the eleventh director of the National Science Foundation (NSF), 1998-2004. In her capacity as director, she served as co-chair of the Committee on Science of the National Science and Technology Council. Dr. Colwell has held many advisory positions in the U.S. Government, nonprofit science policy organizations, and private foundations, as well as in the international scientific research community. She is a nationally respected scientist and educator, and has authored or co-authored 17 books and more than 750 scientific publications. She produced the award-winning film, Invisible Seas, and has served on editorial boards of numerous scientific journals. Before going to NSF, Colwell was President of the University of Maryland Biotechnology Institute and Professor of Microbiology and Biotechnology at the University Maryland. She was also a member of the National Science Board from 1984 to 1990. Dr. Colwell has previously served as Chairman of the Board of Governors of the American Academy of Microbiology and also as President of the American Association for the Advancement of Science (AAAS), the Washington Academy of Sciences, the American Society for Microbiology, the Sigma Xi National Science Honorary Society, and the International Union of Microbiological Societies. She is a member of the National Academy of Sciences, the Royal Swedish Academy of Sciences, Stockholm, the Royal Society of Canada, and the American Academy of Arts and Sciences, and the American Philosophical Society. She is immediate past-president of the American Institute of Biological Sciences (AIBS). Dr. Colwell has also been awarded 55 honorary degrees from institutions of higher education and received numerous awards. Born in Beverly, Massachusetts, Dr. Colwell holds a B.S. in Bacteriology and an M.S. in Genetics, from Purdue University, and a Ph.D. in Oceanography from the University of Washington.

Alice Agogino (NAE)*
Alice Agogino is the Roscoe and Elizabeth Hughes Professor of Mechanical Engineering and an affiliated faculty at the University of California, Berkeley (UCB) Haas School of Business, Energy Resources Group and Studies in Engineering, Science and Mathematics Education. She also directs the Berkeley Expert Systems Technology Laboratory and the Berkeley Instructional Technology Studio. She has served in a number of administrative positions at UCB, including associate dean of engineering and faculty assistant to the executive vice chancellor and provost in educational development and technology. Dr. Agogino continues as principal investigator for the National Engineering Education Delivery System and the digital libraries of courseware in science, mathematics, engineering, and technology. She received a B.S. in mechanical engineering from the University of New Mexico (1975), an M.S. in mechanical engineering (1978) from the UCB, and Ph.D. from the Department of Engineering-Economic Systems at Stanford University (1984). Dr. Agogino is a fellow of the Association of Women in Science and the American Society of Mechanical Engineers. She was elected to National Academy of Engineering (NAE) in 1997 and awarded the NSF Director's Award for Distinguished Teaching Scholars in 2004. She formally served as a member of the Committee on Science, Engineering, and Public Policy Committee on Women in Academic Science and Engineering. She is currently a council member of NAE.

Joan W. Bennett (NAS)*
Joan W. Bennett is a professor in Department of Plant Biology and Pathology and the associate vice president for The Office for Promotion of Women in Science, Engineering and Mathematics at Rutgers University. She is a past president of the American Society for Microbiology and a member of the National Academy of Sciences (NAS). Dr. Bennett has done work in fungal genetics as well as in women's studies. She taught a popular course Biology of Women beginning in 1976 while she was at Tulane University (1971-2006). She is currently a leader of her institution's NSF ADVANCE project on women faculty. Dr. Bennett earned a bachelor's degree in biology and history from Upsala College, and a master's and doctorate degree in botany from the University of Chicago.

Jeremy M. Berg (IOM)*
Jeremy M. Berg became director of the National Institute of General Medical Sciences (NIGMS) in November 2003. He oversaw a $2 billion budget that funds basic research in the areas of cell biology, biophysics, genetics, developmental biology, pharmacology, physiology, biological chemistry, bioinformatics and computational biology. Prior to his appointment as NIGMS director, Dr. Berg directed the Institute for Basic Biomedical Sciences at the Johns Hopkins University School of Medicine in Baltimore, Maryland, where he also served as professor and director of the department of biophysics and biophysical chemistry. In addition, he directed the Markey Center for Macromolecular Structure and Function and co-directed the W.M. Keck Center for the Rational Design of Biologically Active Molecules at the university. Dr. Berg received B.S. and M.S. degrees in chemistry from Stanford University in 1980 and a Ph.D. in chemistry from Harvard University in 1985.

Vivian W. Pinn (IOM)*

Vivian W. Pinn is the former Director of the Office of Research on Women's Health in National Institutes of Health (NIH). She was the first full-time Director of the Office of Research on Women's Health in the Office of the Director of NIH, an appointment she has held since 1991. She was also the NIH associate director for Research on Women's Health. Dr. Pinn came to NIH from Howard University College of Medicine in Washington, D.C., where she had been professor and chair of the department of pathology, and she has previously held appointments at Tufts University School of Medicine and Harvard Medical School. One of her major efforts has been to raise the perception of the scientific community about the importance of sex and gender factors in basic science, clinical research, health care and public policy. She also is currently co-chair, along with the Director of NIH, of The NIH Working Group on Women in Biomedical Careers. Dr. Pinn earned her B.A. from Wellesley College and received her M.D. from the University of Virginia School of Medicine in 1967, where she was the only woman and minority in her class. She returned to Massachusetts to complete her postgraduate training as a Research Fellow in pathology at Massachusetts General Hospital, during which time she also served as Teaching Fellow at Harvard Medical School. Dr. Pinn then joined the faculty of Tufts University School of Medicine and Tufts-New England Medical Center Hospital in 1970. In 1982, when she moved to Howard University, she became the third woman to chair an academic department of pathology in the United States. She is a member of many professional and scientific organizations, in which she held many positions of leadership. She also served as the 88th president (and second woman president) of the National Medical Association from 1989 to 1990. Dr. Pinn has received numerous honors, awards, and recognitions and has been granted 10 honorary degrees of laws and science since 1992. She is a fellow of the American Academy of Arts and Sciences and was elected to Institute of Medicine in 1995.

Patricia Taboada-Serrano

Patricia Taboada-Serrano is assistant professor of chemical and biological engineering at Rochester Institute of Technology. She was born in Brazil and raised in Bolivia. She is a chemical engineer, has a M.Sc. in chemical engineering and a Ph.D. in environmental engineering. She has worked as a research and development engineer at the Center for Applied Research in Bolivia, a postdoctoral research associate at Oak Ridge National Laboratory, and an instructor at Simon Bolivar University (Venezuela), and the Catholic University (Bolivia). She has more than 20 scientific publications in peer-reviewed journals and conference proceedings, numerous conference presentations, and two patents pending. Her research interests include nanothermodynamics and the application of nanotechnology in alternative energy systems. She is a member of the American Chemical Society, the American Institute of Chemical Engineers, the Bolivian Institute of Engineers, and the InterAmerican Network of Academies of Sciences (IANAS) Women for Science Working Group.

Lydia Villa-Komaroff

Lydia Villa-Komaroff is a member of the Board of Directors and the chief scientific officer at Cytonome/ST, LLC. She is also a member of the Board of Directors of the Massachusetts Life Sciences Center. During her 20 year research career, Dr. Villa-Komaroff held positions at Massachusetts Institute of Technology (MIT), Harvard University, University of Massachusetts Medical School and Harvard Medical School. Her research was focused on the molecular biology of protein synthesis, protein processing, and developmental neuroscience. As a science administrator, she has been vice president for research at Northwestern University in Illinois and the vice president for research and chief operating officer of Whitehead Institute for Biomedical Research in Cambridge, Masschussetts. Dr. Villa-Komaroff has served on several National Research Council committees. She is a current member of Committee on Women in Science, Engineering and Medicine and was a member of the Committee on U.S. Competitiveness: Underrepresented Groups and Expansion of the Science and Engineering Workforce Pipeline. She was elected to a 4-year term on the Board of Directors of AAAS and was non-executive chair of the Board of Directors of Transkaryotic Therapies. She is a founding member of the Society for the Advancement of Chicanos and Native Americans in Science and has served as both a board member and vice president of the organization. Dr. Villa-Komaroff received her A.B. from Goucher College and her Ph.D. from MIT.

Susan Wessler (NAS)*

Susan Wessler is distinguished professor of genetics in the Department of Botany & Plant Sciences at the University of California, Riverside. Born in New York City, Dr. Wessler earned her bachelor's degree in biology with honors from the State University of New York at Stony Brook in 1974. She received her Ph.D. in biochemistry from Cornell University in 1980, and was a postdoctoral fellow of the American Cancer Society at the Carnegie Institution from 1980-1982. From 1983-2010, Wessler was in the Department of Plant Biology at the University of Georgia at Athens, where she was assistant, associate, full professor and finally regents professor. Dr. Wessler was elected to membership in the NAS in 1998 and was elected in 2004 to the Council of the National Academies. She was elected as NAS Home Secretary in 2011. She is a fellow of AAAS and the American Academy of Arts and Sciences. She is the recipient of Distinguished Scientist Award (2007) from the Southeastern Universities Research Association, the Stephen Hales Prize (2010) from the American Association of Plant Biology and the Federation of American Societies for Experimental Biology Excellence in Science Award (2012). Her scientific interest focuses on the subject of plant transposable elements and the evolution of plant genomes.

*Denotes members of the National Academy of Science (NAS), National Academy of Engineering (NAE), and Institute of Medicine (IOM).

APPENDIX C

SPEAKER BIOGRAPHIES

Judith C. Giordan
Judith C. Giordan is a partner at ecosVC, a venture development and investment firm. She is a senior advisor to the National Collegiate Inventors and Innovators Alliance, and a member of the board of directors or advisors of start-up companies. Giordan has held executive and leadership positions in R&D and operations spanning a 25-plus-year career.

Her previous executive positions include: vice president and global corporate director of research and development at International Flavors and Fragrances, Inc.; vice-president worldwide research and development for the Pepsi-Cola Company, the global beverage arm of PepsiCo, Inc.; vice president research and development, Henkel Corporation, the North American operating unit of the Henkel Group, and co-founder and managing partner of 1EXECStreet, a successful San Francisco based boutique executive search firm. She has also held management and technical contributor positions at Polaroid and ALCOA.

In addition to her business and university responsibilities, she is active in academic, professional, and industrial organizations. Current and previous positions include: member of the Board of Directors of the American Chemical Society (ACS), the Industrial Research Institute, and the Educational Foundation of the Commercial Development and Marketing Association; member of the Conference Board Advisory Board for Technology Conferences; member of the Board on Chemical Sciences and Technology of the National Research Council; member of the Math and Physical Sciences Advisory Board; Member and Chair of the Waterman Award Committee of the National Science Foundation (NSF); and, chair of the Education and Outreach Committee of the Intangible Asset Finance Society.

Dr. Giordan has held visiting and adjunct professorships at North Carolina State University, Rutgers University and Dartmouth College, and has served as a member of the Board of Advisors of the University of Maryland, College of Life Sciences and the Institute for Strategic Business Markets at Penn State's Smeal Business School. Her research interests and grants focus on two main areas: mechanisms to support and foster women and diversity in science, technology, engineering and math (STEM) and in facilitating STEM intensive entrepreneurship, business and economic development. She is and has been a lead investigator on several NSF grants in these areas.

She is the author of over 200 articles, presentations, and seminars in the areas of entrepreneurship, career development and leadership, intellectual property monetization, market and operational strategy development and implementation, diversity, polymer chemistry, flavor and fragrance technology, and electron spectroscopy. She contributes articles and editorials to magazines and journals including *Research and Technology Management, e-Plant, Chemical Specialties* and numerous international technical journals and web sites. In addition, aspects of her work and activities have been featured in publications including *Working Woman, Chemical*

Week, Chemical Specialties, and *Chemical and Engineering News.* She has also been included in numerous internationally- and nationally-based Who's Who Publications, as well as books, studies and articles on topics including women and diversity, technology, and career development.

Dr. Giordan received her bachelor's degree from Rutgers University (environmental science), her Ph.D. from the University of Maryland (Chemistry), and was an Alexander von Humboldt Post Doctoral Research Fellow at the University of Frankfurt in Germany. She is the recipient of the 2010 ACS Garvan-Olin Medal of the American Chemical Society.

Judy Heyboer
Judy Heyboer has spent the past 27 years in Human Resources (HR), retiring from full-time corporate life in 2000 as senior vice president at Genentech, Inc. in South San Francisco, California. At Genentech, she revamped and re-energized the HR function, and managed the introduction of the entire span of new programs following the company's public offering in 1999. During her tenure, Genentech achieved its initial recognition by *Fortune* Magazine as one of America's "100 Best Places to Work."

Prior to Genentech, Ms. Heyboer spent thirteen years at Acuson Corporation, which she joined in the startup phase to create the HR function and develop the culture and "feel" of the company. She began her human resources career at Spectra-Physics, Inc. where she spent in-depth time in each of the classic HR functions.

In addition to her consulting work in human resources, she devotes a substantial portion of her time and energy to writing, mentoring, advisory work, and making a difference in the community. She continues to be actively engaged in the human resource community, serving as consultant, mentor, speaker, and executive coach. Ms. Heyboer has an MBA from Santa Clara University, and B.A. and M.A. degrees from the University of Michigan.

In 2007, her accomplishments were recognized by induction into the San Mateo County Women's Hall of Fame. She currently serves as a trustee of the Keck Graduate Institute, a graduate school providing advanced degrees in Applied Life Sciences. She also serves on the boards of The Health Trust, Friends for Youth, Resource Area for Teaching, and the Advisory Board of Facing History and Ourselves. She is a senior fellow of the American Leadership Forum.

Manwai (Candy) Ku
Manwai C. Ku is research scientist and program officer in the Office of Diversity and Leadership at Stanford University School of Medicine. She received her B.A. in Chemistry and Sociology from University of Pennsylvania, and her M.A. and Ph.D. in Sociology from Stanford University. Her research interests revolve around issues of gender, race, and diversity in work and occupations. In addition to her work on gender in entrepreneurship, she has published articles on gender and specialty choice in the medical students' program choices. She is currently involved in National Institutes of Health-funded projects to study women's career advancement in academic medicine, as well as team development and processes in collaborative science research.

Gail K. Naughton
Gail K. Naughton, Ph.D., has been the dean of the College of Business Administration at San Diego State University (SDSU) since August 2002. Prior to that she spent more than 15 years at

Advanced Tissue Sciences, where she was the company's co-founder and co-inventor of its core technology. While at SDSU, Dr. Naughton spearheaded a number of unique MBA programs in partnership with industry, played an instrumental role in the industry committee of the Association to Advance Collegiate Schools of Business (AACSB) and served as a member of the Board of Directors of AACSB International. She has spent more than 15 years extensively researching the tissue engineering process, holds more than 90 U.S. and foreign patents and has been extensively published in the field.

In 2000, Naughton received the 27th Annual National Inventor of the Year award by the Intellectual Property Owners Association in honor of her pioneering work in the field of tissue engineering. Naughton sits on the Board of Directors of Celera (NASDAQ:CRA) and CR Bard (NYSE: BCR).

Sharon Nunes
Sharon Nunes is vice president of Global Government and Smarter Cities Strategy & Solutions, where she directs the strategy & integration of solutions for state, local, and national governments, including innovations for IBM's Smarter Planet initiative. In 2009, she was a Women's History Month Honoree for being one of the "Women Taking the Lead to Save Our Planet," and in June 2009, she was inducted into the Women in Technology International (WITI) Hall of Fame. Dr. Nunes received two awards for her mentoring of technical women: IBM's 2004 Fran Allen Mentoring Award and National Association for Female Executives's 2006 Women of Excellence national award. She was a National Academy of Engineering (NAE) Frontiers of Engineering fellow in 2000 and a member of the NAE "Engineer of 2020" advisory board. Dr. Nunes received her Ph.D. in materials science from the University of Connecticut.

E. J. Reddy
As a research fellow, E. J. Reedy oversees grants and conducts academic and policy research for the Ewing Marion Kauffman Foundation in the field of entrepreneurship. He has been significantly involved in the coordination of the Kauffman Foundation's entrepreneurship and innovation data-related initiatives. He is a co-principal investigator on the Kauffman Firm Survey, an eight-year longitudinal study of new businesses, and the Foundation's multi-year series of symposiums on data, as well as many web-related projects and initiatives. Mr. Reedy joined the Kauffman Foundation in 2003. Prior to joining the Kauffman Foundation, Mr. Reedy was a senior analyst at the Federal Reserve Bank of Kansas City and had extensive experience in non-profit management. Previously, he was financial director and executive co-director of the Center for Community Outreach at the University of Kansas in Lawrence.

Mr. Reedy is active in several civic initiatives, and has been the recipient of the Founders' Award from the Lesbian and Gay Community Center of Greater Kansas City. Mr. Reedy earned his bachelor's degree, Phi Beta Kappa, in economics, mathematics, and American studies from the University of Kansas.

Lucinda Sanders
Lucinda Sanders is CEO and co-founder of the National Center for Women & Information Technology (NCWIT), a consortium of more than 300 corporations, universities, and non-profits working to increase the participation of girls and women in computing and information technology. She also serves as executive-in-residence for the ATLAS Institute at the University of Colorado at Boulder.

Ms. Sanders has an extensive industry background, having worked in R&D and executive positions at AT&T Bell Labs, Lucent Bell Labs, and Avaya Labs for over 20 years, where she specialized in systems-level software and solutions (multi-media communication and customer relationship management). In 1996, Ms. Sanders was awarded the Bell Labs Fellow Award, the highest technical accomplishment bestowed at the company, and she has six patents in the communications technology area.

Ms. Sanders serves on several high-tech startup and non-profit boards, and frequently advises young technology companies. She has served on the Mathematical Sciences Research Institute Board of Trustees at the University of California at Berkeley; as well as on the Information Technology Research and Development Ecosystem Commission for the National Academies.

In 2004, she was awarded the Distinguished Alumni Award from the Department of Engineering at CU and in 2011 she was recognized with the university's George Norlin Distinguished Service Award. She has been inducted into the WITI Hall of Fame and was named by the U.S. Secretary of Commerce to serve on the department's Innovation Advisory Board. Ms. Sanders received her B.S. and M.S. in Computer Science from Louisiana State University and the University of Colorado at Boulder, respectively.

Maxine L. Savitz
Maxine L. Savitz retired as the general manager for technology partnerships at Honeywell, Inc. Her areas of expertise include energy efficiency R&D and products in the transportation, industry, and buildings sectors; aerospace technology; and, integration of R&D between laboratories and business units. During her career at Honeywell, she oversaw the development and manufacturing of innovative materials for the aerospace, transportation, and industrial sectors.

Dr. Savitz serves as a member of the President's Council of Advisors on Science and Technology. She is the former deputy assistant secretary for Conservation at the Department of Energy (DOE) from 1979 to 1983. She received the Outstanding Service Medal from the Department of Energy in 1981. Prior to her DOE service, she was program manager for Research Applied to National Needs at NSF. Following her government service, Dr. Savitz served in executive positions in the private sector, including: President of Lighting Research Institute, assistant to the vice president for engineering at the Garrett Corporation, and General Manager of Allied Signal Ceramic Components.

Dr. Savitz currently serves as vice president of the National Academy of Engineering (NAE) and is a member of AAAS. She was appointed to the National Science Board in 1998-2004. She is a member of Advisory Boards at Sandia and Pacific Northwest National Laboratories and Sandia National Laboratory. She had been a member of the Secretary of Energy Advisory Board, the Laboratory Operations Board and advisory committees at Oak Ridge National Laboratory. She serves on the board of directors of the American Council for an Energy Efficient Economy and Draper Laboratory. She had previously served on the board of directors of the Electric Power Research Institute and the Energy Foundation. Dr. Savitz received a B.A. in Chemistry from Bryn Mawr College and a Ph.D. in Organic Chemistry from Massachusetts Institute of Technology.

Sheldon M. Schuster

Sheldon M. Schuster is currently president of Keck Graduate Institute of Applied Life Sciences (KGI) in Claremont, California. A member of the Claremont College Consortium, KGI features a professional graduate program (Master of Bioscience) designed to educate scientists and engineers to be the bioscience business leaders of the 21st century. One of the first professional science masters degrees in the United States, the MBS emphasizes interdisciplinary and team-based active learning, and has become the premier degree for those wishing to enter and lead the biotechnology, medical device, pharmaceutical development and modern agricultural industries. Schuster has also expanded the program to include new centers focused on bioprocessing, rare diseases, and biomarkers in order to expand education and research in these critical areas affecting not only industry but the entire heath care system.

Prior to joining KGI, Dr. Schuster was the interim assistant vice president for research, director of The Biotechnology Program and professor of biochemistry and molecular biology at the University of Florida. His research focused on the mechanism of tumor drug resistance and the rational design of potential anti-tumor therapies based on studies of specific enzyme structures. In addition, he initiated a research program attempting to use novel gene analysis tools to determine the microbial etiology of numerous chronic human diseases. His research work has been funded by the National Cancer Institute and the American Cancer Society, and has resulted in over 140 peer-reviewed publications and ten patents.

Dr. Schuster is a graduate of the University of California at Davis and he earned his Ph.D. from the University of Arizona. He has worked at the Institute for Enzyme Research at the University of Wisconsin and was on the faculty at the University of Nebraska as Professor of Chemistry and Biological Sciences. Dr. Schuster joined the faculty of the University of Florida in 1989 and moved to KGI in 2003. He is active in the development of start-up companies from KGI and technologies from other universities. Dr. Schuster was one of the founders of Restoragen, Inc. (formerly BioNebraska Inc.), a biotechnology company that produced recombinant peptide therapeutics for the cure of diabetes and osteoporosis, and AquaGene, a company developing technologies to produce biopharmaceuticals using fish as the bioreactor.

Dr. Schuster has had a long-standing interest in biotechnology education and is active nationally and locally promoting development of the workforce necessary to sustain a viable biotechnology sector. He has been active in leadership roles in BIO and the Council for Biotechnology Centers, and is presently associate editor for biotechnology of the journal *Biochemistry and Molecular Biology Education*. He has been exploring the best methods to bridge the gap from basic research ongoing in universities to the applied technologies required for commercial development.

Dr. Schuster serves as a member of the accrediting commission for Senior Colleges and Universities of the Western Association of Schools and Colleges, an accrediting agency for California, Hawaii, and several U.S. territories. Additionally, he was recently elected as a Fellow of AAAS.

Caroline Simard

Caroline Simard is associate director of Diversity and Leadership at the Stanford University School of Medicine. In this role, she leads the implementation of new initiatives around faculty career flexibility. Previously, she directed the Anita Borg Institute's (ABI) research initiatives. She led the design, data collection and analysis, writing, and dissemination of the Institute's major research initiative: "Climbing the Technical Ladder: Obstacles and Solutions for Mid-

Level Women in Technology," which has received national media attention. She spearheaded executive engagement programs directed at supporting organizational change for greater retention and advancement of technical women. She is a frequent speaker on organizational and individual strategies for talent management in academia and industry. Dr. Simard is passionate about social science research and its role in creating practical solutions to social problems.

Prior to ABI, Dr. Simard was a researcher at the Center for Social Innovation of the Stanford Graduate School of Business. She holds a Ph.D. in communication studies from Stanford University, with a focus on organizational theory, high-technology industries, and social networks. She holds a Bachelor's Degree from Université de Montréal and a Masters Degree in Communication and Information Studies from Rutgers University. Dr. Simard's publications have focused on gender and technical human capital, the barriers to the diffusion of best practices, managing open innovation, regional clusters of innovation, and social networks. She serves on the Leadership Team of the NCWIT and is a former member of the editorial committee at the *Stanford Social Innovation Review*.

Laurel Smith-Doerr
Laurel Smith-Doerr is associate professor of sociology at Boston University. Her book *Women's Work: Gender Equality vs. Hierarchy in the Life Sciences* (2004, Boulder, CO: Lynne Rienner Publishers) explains how network forms of organization are more conducive to gender equity than are more rule-bound hierarchical settings. Dr. Smith-Doerr has published in the areas of organizational and economic sociology, science and society, and gender and work. Her published research has appeared in a variety of journals and edited volumes including *Administrative Science Quarterly, Gender & Society, Handbook of Economic Sociology, Handbook of Science and Technology Studies, Journal of Technology Transfer, Minerva, Regional Studies,* and *Sociological Forum*. Dr. Smith-Doerr's research has examined tensions in the institutionalization of science, including examination of networks in the biotechnology industry, commercialization in the university, contributions of immigrant entrepreneurs, gendered organizations, and scientists' responses to ethics education requirements. From 2007 to 2009, she was program director of Science, Technology & Society at NSF. She received the NSF Director's Award for Collaborative Integration for her work as Chair of Cross-NSF program in Ethics Education in Science and Engineering, and on the America COMPETES Act policy committee.

Michael S. Teitelbaum
Michael S. Teitelbaum is a senior advisor to the Alfred P. Sloan Foundation, where he advises the Foundation on the management of its Professional Science Master's program. From 1983 to 2010, Dr. Teitelbaum served as a program director for the Foundation and was responsible for overseeing a number of grantmaking programs, including the Sloan Research Fellowships, the Professional Science Master's program, the Science and Engineering Workforce program, the Federal Statistics program, the Sloan Public Service Awards, and the Sloan Awards for Excellence in Teaching Science and Mathematics. His research interests include the causes and consequences of very low fertility rates; the processes and implications of international migration; and patterns and trends in science and engineering labor markets in the U.S. and elsewhere. He is the author or editor of 10 books and a large number of articles on these subjects. Previously he was a faculty member at Princeton and Oxford Universities, and served

as vice chair and acting chair of the U.S. Commission on International Migration. He was educated at Reed College and at Oxford University, where he was a Rhodes Scholar.

Jessica Townsend
Jessica Townsend is an assistant professor of mechanical engineering at the Franklin W. Olin College of Engineering in Needham, MA. Olin College is a new undergraduate engineering institution with a focus on design and entrepreneurship. Dr. Townsend teaches several of the mechanical engineering core courses at and also advises student design teams in Olin's Senior Capstone program. Her current research interests include development and characterization of engineered fluids (nanofluids) for electronics cooling applications. Dr. Townsend received her Ph.D. in aeronautics and astronautics from MIT, where she developed evaporatively cooled turbine blades for advanced aircraft engines. Prior to returning to school for her doctorate, she spent three years in industry as an aerospace performance engineer at Hamilton Sundstrand Power Systems.

Dr. Townsend received her M.S. from the Mechanical and Aeronautical Engineering Department at the University of California at Davis. She earned her B.S. at the University of Massachusetts at Amherst in mechanical engineering and credits her mentors at UMass for their early influence in pointing her towards a career in engineering education.

She was also the recipient of the American Institute of Aeronautics and Astronautics (AIAA) Foundation Wilbur and Orville Wright Graduate Research Award, the AIAA Foundation Gordon C. Oates Air Breathing Propulsion Graduate Award and the American Association of University Women Engineering Dissertation Fellowship.

Barbara Wallner
Barbara Wallner has recently retired and serves currently as consultant to several biotechnology companies. Before her retirement she held the positions of senior vice president for Technology Operations and chief technology officer at ZIOPHARM Oncology, Inc, a publically traded company that is developing cancer therapeutics. She was also founder, president and chief executive officer of Chymic Therapeutics, Inc., a start-up company that was developing therapeutics for cancer, autoimmune and infectious diseases.

Before joining ZIOPHARM Oncology, Dr. Wallner was senior vice president of Research and chief scientific officer at BioTransplant, Inc. Previous to that she was a founder, senior vice president for research and chief scientific officer of Point Therapeutics (acquired by DARA BioSciences); before that she was vice president for Research at ImmuLogic, Inc. and held several management positions at Biogen (now Biogen Idec) where she invented Amevive® now marketed for psoriasis by Astellas Pharma. During her 30 years of industry experience, Dr. Wallner has published 68 scientific papers and authored 32 issued U.S. patents. She has served on several Scientific Advisory Boards.

Susan Windam-Bannister
Susan Windam-Bannister is the first president and CEO of the Massachusetts Life Sciences Center, a quasi-public organization charged with administering the 10-year $1 billion life sciences initiative enacted by the Massachusetts Legislature in June 2008. The Life Sciences Center is the hub for all sectors of the state's life sciences community—biotechnology, pharmaceuticals, medical devices, medical diagnostics and bioinformatics.

Since assuming the executive leadership of the Life Sciences Center in July 2008, Dr. Windham-Bannister has been responsible for the overall implementation of the life sciences initiative, including staffing the center, developing policies and procedures, creating a brand, and formulating the investment strategy. The center's portfolio of investment is promoting economic development, catalyzing innovation, strengthening Massachusetts' global leadership position in the life sciences, and accelerating the commercialization of promising treatments, therapies and cures. Under Windham-Bannister's leadership, in just three years the center has invested $218 million, leveraged another $700 million in matching investment capital, and created over 7,000 new life sciences jobs in the commonwealth.

Before assuming her role at the Life Sciences Center, Dr. Windham-Bannister was a founding partner of Abt Bio-Pharma Solutions (ABS), a boutique consulting firm serving life sciences companies. Within ABS, Windham-Bannister managed the Commercial Strategy Group. In her 35 year consulting career, she has been instrumental in the successful launch of a number of well-known therapeutics, medical devices, and novel biomarkers.

Dr. Windham-Bannister has co-authored two books: *Competitive Strategy for Health Care Organizations* and *Medicaid and Other Experiments in State Health Policy*. She also has written numerous articles on competition in today's health care marketplace. She holds a B.A. from Wellesley College, a doctorate in health policy and management from the Heller School at Brandeis University, and was a post-doctoral fellow at Harvard University's John F. Kennedy School. She completed her doctoral work under a fellowship from the Ford Foundation.

APPENDIX D

WORKSHOP PARTICIPANTS

(Participants are listed alphabetically with affiliations at the time of the workshop.)

Tina Abdollah
Senior Architect
IBM

Alice Agogino
Roscoe and Elizabeth Hughes Professor of Mechanical Engineering
University of California, Berkeley

Dilek Alkaya
Advanced Control Engineer
Tesoro Golden Eagle Refinery

JoAnn Armenta
Executive Director
Purpose Focused Alternative Learning Corporation

Florence Bonner
Senior Vice President for Research & Compliance
Howard University

Rebecca Breitenkamp
Chief Financial Officer
Intezyne Technologies

Carrie Brubaker
Mirzayan Fellow
The National Academies

Jessica Christenson
Graduate Student
City of Hope

Kendra Clark-Branton
Project Engineer
VF Corporation

Catherine Didion
Director, Committee on Women in Science, Engineering, and Medicine
The National Academies

Manqing Ding

Catherine Domier
University of California, Los Angeles

Baat Enosh
Entrepreneurial Alliance Program Manager
National Center for Women and IT

Laurimar Escudero
University of California, Irvine

Peace Esonwune
SMB Operations Engineering and Science Research Assistant
Stanford University

Haleh Farahbod
Postdoctoral Fellow
University of California, Los Angeles

Allan Fisher
Senior Vice President
Laureate Education, Inc.

Elizabeth Frayne
President
Frayne Consultants

Lynne Friedmann
Science Writer
Friedmann Communications

Linda Garverick
Principal
Coactive Consultants

Judith Giordan
Senior Advisor
National Collegiate Innovators and
Inventors Alliance

Florence Haseltine
Director, Center for Population Research,
EKS-NICHD, National Institutes of Health

Judy Heyboer
Human Resources Consultant
Genentech Inc.

Nicole Hernandez

April Ho
Graduate Student Researcher
University of California, Los Angeles

Mitra Hooshmand
Graduate Student
University of California, Irvine

Xue Hua
Postdoctoral Fellow
University of California, Los Angeles

Pamela Jeter
Research Specialist
University of California, Irvine

Heather Johnston
Graduate Student
Irell and Manella Graduate School of the

Mary Juhas
Program Director, Comprehensive Equity at
Ohio State
The Ohio State University

Swati Kadam
Graduate Student
City of Hope National Medical Center

Mi Jung Kim
University of California, Los Angeles

Monica Kim
University of California, Irvine

Manwai (Candy) Ku
Department of Sociology
Stanford University

Jingyu Li

Sarah Madsen
University of California, Los Angeles

Cecelia McCloy
President/CEO
Integrated Science Solutions, Inc.

Audry McGillicuddy
Technical Account Manager
IBM

Jill McNitt-Gray
Professor
University of Soutern California

Ellis Meng
Assistant Professor
University of Southern California

Carolyn Merry
Professor and Chair
The Ohio State University

APPENDIX D: WORKSHOP PARTICIPANTS

Gail Naughton
Dean
College of Business Administration
San Diego State University

Yiling Nie
Postdoctoral Fellow
University of California, Los Angeles

Sharon L. Nunes
Vice President, Strategic Growth Initiatives
Big Green Innovations
IBM Corporation

Jo Ann Oravec
Associate Professor
University of Wisconsin, Whitewater

Tayhas Palmore
Professor of Engineering
Brown University

Lili Peng
Bioengineering Researcher
University of California, San Diego

Vivian Pinn
Director, Office of Research on Women's Health
National Institutes of Health

Mikki Popovich
Owner
One Oh One Eight Enterprises, Inc.

Raquel Raices
Research Fellow
Beckman Research Institute, City of Hope

Ana De La Ree
Graduate Student Researcher
University of California, Irvine

E. J. Reedy
Manager, Research and Policy
Kauffman Foundation

Heidi Ries
Dean for Research
Air Force Institute of Technology

Amy Robb
Research Opportunities Coordinator
ADVANCE Program, Brown University

Claire Robertson
Graduate Student Researcher
University of California, Irvine

Karla Shepard Rubinger
Executive Director
Rosalind Franklin Society

Pardis Sabeti
Assistant Professor
Harvard University
Broad Institute

Roya Saleh
Senior Physician-Scientist
University of California, Los Angeles

Lucinda Sanders
CEO and Co-founder
National Center for Women & Information Technology and the University of Colorado

Miruna Sasu
PhD Candidate Biology/Statistics
Pennsylvania State University

Maxine L. Savitz
General Manager, Technology/Partnerships
Honeywell Inc. (Retired)

Jean Schelhorn
Associate Vice President,
Technology Licensing & Commercialization

Sheldon Schuster
President
Keck Graduate Institute of Applied Life Sciences

Andrea Schweitzer
U.S. Project Manager
Int. Year of Astronomy

Caroline Simard
Director of Research and Executive Programs
Anita Borg Institute for Women and Technology

Laurel Smith-Doerr
Associate Professor of Sociology
Department of Sociology
Boston University

Ruthlyn Sodano
Postdoctoral Fellow
Integrated Substance Abuse Programs
University of California, Los Angeles

Kathleen Spencer
Ph.D. Candidate
California Institute of Technology

Marinela Stack
Electronics Design and Project Engineer

Adele Tamboli
California Institute of Technology

Anil Tarachandani

Michael Teitelbaum
Program Director
Alfred P. Sloan Foundation

Jessica Townsend
Assistant Professor of Mechanical Engineering
Franklin W. Olin College of Engineering

Becky Tsai
University of California, Irvine

Katherine Tsai
Stanford University

Moran Valensi-Kurtz

Carolyn Vallas
Director, Center for Diversity in Engineering
School of Engineering and Applied Science, University of Virginia

Lydia Villa-Komaroff
Chief Scientific Officer
Cytonome/ ST, LLC

Barbara Wallner
President and CEO
Chymic Therapeutics, Inc.

Susan Wessler
University of Georgia Foundation Chair
Biological Sciences, University of Georgia

Susan Windham-Bannister
President/CEO
Massachusetts Life Sciences Center

Elizabeth WoldeMussie

Lilian Wu
Program Executive
IBM Global University Programs

Shanshan Xu
University of California, Irvine

Yajie Yang
IBM

Noriko Yokoyama
University of California, Irvine

Sohila Zadra